The Exploration of the Universe

by Giorgio Abetti

THE SUN *Translated by J. B. Sidgwick*
STARS AND PLANETS *Translated by V. Barocas*

with Margherita Hack
NEBULAE AND GALAXIES *Translated by V. Barocas*

Photograph taken by Lunar Orbiter IV at an altitude of 1690 miles from the Moon's surface. The dark irregular area is Mare Veris (spring sea), that lies at the foot of the Rook Mountains. Photo by courtesy of NASA

The Exploration

of the Universe

GIORGIO ABETTI

Translated by V. Barocas

FABER AND FABER LIMITED
London: 24 Russell Square

First published in 1965
as Esplorazione dell'universo
by Laterza and Figli, Bari
this translation first published in mcmlxviii
by Faber and Faber Limited
London: 24 Russell Square W.C.1
Made and printed in Great Britain
by William Clowes and Sons, Limited
London and Beccles

S.B.N. 571 08688 8

Contents

CONTENTS

Plates

Photograph taken by Lunar Orbiter IV at an altitude of 1690 miles from the Moon's surface. (Photo by courtesy of NASA)

frontispiece

between pages 32 and 33

1. Sunspots: photograph taken by Stratoscope I. (Project Stratoscope, Princeton University, sponsored by NSF, ONR and NASA)
2. The 200-inch Hale telescope Mt. Palomar
3. Great Nebula in Andromeda M.31–NGC 224. Satellite nebulae NGC 205 and 221 also shown. 48-inch Schmidt Mt. Palomar
4. Cluster of galaxies in Coma Berenices. 200-inch Mt. Palomar
5. (a) Exterior of the horizontal Snow telescope, together with the two solar towers at Mt. Wilson, California
   (b) The solar tower at Arcetri Observatory, Florence
6. The 250-ft radio telescope at Jodrell Bank (Manchester University)
7. (a) Hurricane of September 21st, 1948, at 11 h 31 min E.S.T.
   (b) The same hurricane at 16 h 10 min E.S.T.

between pages 144 and 145

8. Spectroheliogram obtained in the hydrogen line Hα with a Lyot filter at the Sacramento Peak Observatory (U.S.A.)
9. Spectroheliogram of flares, filaments, spots, taken with the hydrogen line Hα November 28, 1958, 13 h 55 min U.T. at the Anacapri station of the Fraunhofer Institute
10. (a) Photograph of a prominence in the light of Hα taken with a Lyot filter, June 12th, 1937, at 13 h 28 min U.T. (Meudon Observatory)

7

10. (b) Eruptive prominence taken in the light of Hα at the Sacramento Peak Observatory
11. The corona of June 19th, 1936. Polar type. Italian expedition to Sara, U.S.S.R.
12. The corona of February 25th, 1952. Intermediate type. Italian expedition to Khartoum, Sudan
13. a, b, c, Various types of aurorae
14. Photograph taken by Lunar Orbiter II
15. The Moon at first quarter (Moore and Chappell) Lick Observatory
16. The other side of the Moon—wide angle view of the far side of the Moon received from Lunar Orbiter V, showing surface features as small as 500 metres across
17. Photograph of Mars taken by Mariner 4, with green filter, and at a distance of 7,800 miles from the surface of the planet, showing Atlantis between Mare Sirenum and Mare Cimmerium
18. Comet Arend-Roland (1956). Arcetri Observatory

Translator's Note

Professor Abetti has written many books on astronomy that are well known and have been widely used as reference books in Italy. Several of them have also been translated into various languages.

Apart from his more classic works, Professor Abetti, for many years, has been known to a large public in Italy for his non-technical and popular writings aimed at increasing the general interest in astronomy. This valuable work was continued after his retirement.

Recently, the Italian publishers Laterza decided to collect together a selection of Professor Abetti's articles written throughout the years, and to present them to the public in book form. This is the origin of the present book which consists of a number of articles on various aspects of astronomy, each article being self contained.

The reader will find in some cases inevitable duplications and repetitions. At the request of Professor Abetti I made some modifications when preparing the book for the English edition. I have tried to avoid making too many changes to the original text. Some new articles dealing with more recent discoveries and astronomical events have been added, together with new diagrams, plates and an index. A bibliography has not been included; it seemed unnecessary on account of the nature of the book.

Jeremiah Horrocks and   V. B.
Wilfred Hall Observatories
Preston
March 1968

Preface

MAN AND THE CONQUEST OF SPACE

The wonderful and rapid progress in the exploration of our own atmosphere and of further parts of space which we witness daily, has led mankind to take a much greater interest than ever before in all celestial events and phenomena.

This interest is shown by the eagerness with which new astronomical discoveries are received and by the increased attention paid to celestial phenomena in general. The desire to know more about the mysterious universe in which we live has grown considerably in recent years. Discoveries made not only by astronomers but also by scientists working in other branches of science with new technical devices, are eagerly followed by many people.

It is extremely difficult at this stage to predict the development of the investigations which are being undertaken and to forecast the results which will be obtained in the near future. One thing is certain: the information which will be obtained will be of great importance. For instance, radio astronomy, one of the youngest branches of astronomy, has already achieved great success and has widened considerably the limits of the universe beyond those formerly reached by the largest optical telescopes available on Earth. Moreover, the possibility of obtaining observations of celestial objects with instruments working above our own atmosphere, of bringing these instruments to within a very short distance of the Moon and some of the planets, and even of landing

11

such equipment on them, is bound to increase our knowledge and lead to discoveries which at this stage we cannot even imagine.

In this book we propose to give a survey of our present knowledge of the new astronomy, of the new methods used and of the results obtained. It is our intention to discuss some of the most important astronomical questions and to address ourselves to those who have a general interest in astronomy, but perhaps have not the necessary background to undertake the serious study of this subject.

Perhaps we ought to point out here how much confusion exists in people's minds when they refer to astronomical problems concerned with the exploration of space by using the common expression 'the conquest of space'. This, probably, is due to the various meanings that can be attributed to the word 'space' and to the various problems which may well be the province of several branches of science. If we limit ourselves to the volume of space occupied by the terrestrial atmosphere, we find a very wide field of research which in reality belongs to that branch of science which we call 'geophysics'. Rockets and artificial satellites, equipped with special instruments, have in recent years greatly contributed towards our understanding of the phenomena which take place in the atmosphere, and in many cases have revealed new facts to us. Recently we have witnessed the wonderful achievements of both the United States and Russia who have succeeded for the first time in placing men in space and allowed them to stay there for several days, revolving around the Earth at a fantastic speed. Even so, these 'flights', so far, have still been within the terrestrial atmosphere. No doubt this is a necessary part of the tests before man can venture further into interplanetary space, in the exploration of that space where the planets move in their revolution around the Sun. When we enter this interplanetary space and, further, when we enter the interstellar space, we shall be able then to talk of observations and questions as belonging to astrophysics.

For the time being, the easiest way to obtain these observations is by means of, first, unmanned and then manned artificial satellites, which, while they remain still relatively near to the Earth, can nevertheless obtain observations from above the terrestrial atmosphere and give us completely new data on the celestial bodies.

A task which is even more difficult has already had an auspicious beginning. We are referring to the photographs of Mars and of the other side of the Moon, to the observations of Venus at close quarters and to the landing of capsules on the surface of the Moon. The next step, as is well known, will consist of the possible landing of astronauts on the Moon.

It is necessary to remember, as we shall try to explain in the following pages, that both in the case of the exploration of interplanetary space and even more in the case of interstellar space, the distances involved are very great indeed, particularly when compared with the short life-span of man who was created to live on Earth and who has many limitations which at present may appear almost insurmountable. Lastly, let us not forget that beyond that 'space' which we have called interstellar and which is within our Galaxy, there exists still an even larger 'space' extending to the galaxies up to distances of several thousand millions of light-years. Not even our largest optical telescopes and radio telescopes can reach the limits of this 'space'.

G.A.

Arcetri, June 1966

1 · Introduction: The New Astronomy

Johannes Kepler, astronomer and mathematician at the court of the emperor Rudolph II in Prague, published his *Astronomia Nova* in 1609. This was really a new astronomy which introduced new revolutionary ideas about the movement of the celestial bodies belonging to the solar system. The planet Mars, which until then had resisted all attempts to explain its apparent erratic movement in the sky, was, in Kepler's flowery language, 'at last a prisoner in chains at the foot of the imperial throne'.

Kepler used the long series of observations made by Tycho Brahe at his observatory 'Uraniborg' on the island of Huen off Denmark and was able to put forward his now famous laws concerning the motion of the planets around the Sun, before Newton expounded the theory of universal gravitation. We can almost say that the wonderful theoretical branch of astronomy which we call 'celestial mechanics' was born at that time. In the centuries which were to follow, this branch of astronomy made such progress as to enable man to predict with great accuracy the position of the Sun, of the Moon and of all planets, not only in the past but also in the present and in the future. This enabled man to predict eclipses and many of the phenomena which occur in the sky. Celestial mechanics having reached a very high degree of development and perfection remained static and almost forgotten in modern times. The reason for this is that perhaps it was felt that celestial mechanics had performed its task. A new era however was to begin for this branch of astronomy in what we may call 'modern astronomy'. The methods so far used were revolutionized while their fundamental principles were still maintained.

For example, when an astronomer has discovered a comet and has observed its course on several nights and has lost it when it approached the Sun, how can he re-locate it if it becomes visible to the power of the telescope he uses? This problem was solved by theoretical astronomers by means of celestial mechanics. From the few observations available of the position of the comet they were able to determine its orbit, that is its track around the Sun. The task was made easier by means of trigonometrical tables and by calculating machines. Nevertheless, it was long and tedious work complicated by the fact that the comet or the planet or the satellite in question, in its journey is subject to perturbations produced by the attraction of other bodies which might happen to be in its vicinity.

Man has succeeded in building 'artificial satellites' which although made on a much smaller scale, yet behave like real satellites. In order to set an artificial satellite in motion, man must not only make the machine and produce the necessary propellent, but he must also know at what velocity the satellite must move and which orbit it must follow in order to go around the Earth or the Sun or approach a planet. In other words, man makes use of celestial mechanics in a new guise. This use sets up many new problems, the solutions of which are made much easier nowadays by the use of modern electronic computers.

Astronomers by consulting the old treatise of celestial mechanics can tell us what must be the exact initial velocity of an artificial satellite so that it can follow a given orbit, as has already been the case for the many artificial satellites which are at present revolving around the Earth. Once the artificial satellite has left the Earth it is followed by means of powerful optical telescopes located at various stations all over the surface of our globe. As long as it is within range photographs are obtained which enable the astronomer to determine the position of the artificial satellite at any particular instant of time. Alternatively, a radio transmitter, carried by the artificial satellite, emits signals which, when received on Earth, enable us to determine the satellite's position. Some of the centres which collect the information, such as the Astrophysical Observatory of the Smithsonian Institute at Cambridge, Massachusetts, collate very quickly all the data received. This data is then prepared according to the laws of celestial mechanics and fed into an electronic computer. In a very short time the machine will do the calculations and supply the elements of the orbit. In the old days a great number of people would have taken a much longer time to do exactly the same thing.

Once the elements of the orbit are obtained, they are used to calculate the 'ephemeris' of the artificial satellite, that is to say the exact position of the satellite in the sky at any time, just as the astronomical ephemeris gives the future positions of the members of the solar system.

The mystery of the Creation still appears miraculous. Let us take our small Earth–Moon system. This consists of the Moon which on an astronomical scale is infinitesimal but which is rather large when compared with terrestrial objects. It continues to revolve around a larger body, our own Earth, for a time which can be considered infinite when measured against the span of human life and is subject to a mysterious force which keeps it in its position without any apparent loss of energy. The same applies to artificial satellites. Man causes the expenditure of a certain amount of energy in order to make them reach a given height or in order to make them leave the gravitational field of the Earth, but then, without appearing to use any other energy, these artificial satellites continue to revolve around the body which attracts them, for a time which again could be considered to be infinite. In actual fact, in the case of artificial satellites which orbit around the Earth, the atmosphere, although extremely rarefied at those heights, still presents a certain amount of friction to the satellite's motion. As a result the satellite is compelled to come closer to the Earth and to accelerate, until friction becomes so great that the satellite burns out before actually falling on the Earth.

By means of satellites launched into orbits which are at various distances from the Earth, it has been possible to discover that our atmosphere extends beyond what was once thought to be the limit. It is true that at these heights the atmosphere is extremely tenuous, but it has revealed certain physical characteristics which so far were unknown.

The motion of artificial satellites is affected not only by air friction but also by perturbations. In the case of natural satellites the perturbations are due to other celestial bodies, but in the case of artificial satellites, the main cause of the perturbations is the shape of the Earth which is only approximately spherical.

The Earth, as is well known, is flattened at the poles. In addition there are various continents with their ranges of mountains and the oceans. The artificial satellite, in its orbital motion around the Earth 'feels', as it were, these irregularities and therefore its regular motion is affected by them. These irregularities, or perturbations, in the motion of the artificial satellite can be detected by the accurate calculations of the astronomers and as a result the value of the flattening of the Earth can be determined, as well as the actual shape of the Earth. In other words, by means of these observations some geodetic problems can be solved. In passing we can note that when we calculate and determine the orbits of artificial satellites we are still in the field of celestial mechanics. In actual fact all the investigations made by means of these satellites and all the valuable results already obtained do not belong to astronomy, but rather to geophysics, the science concerned with the physical characteristics of the Earth.

Although it is not possible to draw a line between sciences which are very similar, there is no doubt that geophysics may be considered to be a form of experimental astronomy, since from the study of a celestial body, such as the Earth, we can extend the results to other celestial bodies similar to the Earth. Moreover, the field of research opened to astronomy by the use of artificial satellites is indeed very wide. One of the most important facts is that the satellites can operate beyond the denser and lower layers of our atmosphere, and can therefore explore the sky from a position beyond these layers. The importance of this lies in the fact that these layers of the terrestrial atmosphere not only absorb a great deal of the radiation passing through them, but also tend to introduce distortions in that part of the radiation which succeeds in reaching us. Already it has been possible to obtain photographs of the Sun from a height of about 18 miles above sea level. These photographs (Plate 1) which were obtained by means of telescopes carried by balloons, have already given us a better knowledge of the turbulence of the solar surface which appears to be in a state of constant agitation. By means of rockets carrying special instruments, astronomers have already obtained spectra of the Sun in the ultraviolet region from a height of approximately 240 miles above the Earth. These spectra cannot be obtained from the Earth, since the ultraviolet radiation does not penetrate our atmosphere. Much more important information is obtained by means of space probes such as those which have photographed the invisible side of the Moon, or those which have photographed Mars from a close range. The recent landing on the Moon of capsules which are able to send close-up photographs of the surface of the Moon will advance our knowledge of the surface and the physical conditions of our natural satellite.

So far all these questions have had to be investigated by means of our terrestrial telescopes (fig. 1) and the study of light which filters through our very disturbed atmosphere. Even if unmanned stations will not be able to soft-land on the major planets such as Jupiter and Saturn, because they are liquid and gaseous, nevertheless space probes which will succeed in getting close to them, will certainly give us enough data to help us understand better their physical conditions.

Even if we are very optimistic about interplanetary explorations we must, however, remember that the astronomers are aiming at goals which are far beyond those attainable by spacecraft. The Sun, which plays such an important part in our lives, is within reach of man but man will not be able to approach it because of the enormous energy and heat which it radiates. The same applies, and to a much greater degree, to all the stars which we see in the sky. It is true that some of them may not radiate as intensely as the Sun, but their distances from us are so great that man will not be able to cover such distances even if travelling at the speed of 186,000 miles per second for the whole of his comparatively short life. Having said all this we must, however, admit that at present we have gained so much knowledge about the physical constitution of the stars, that in many

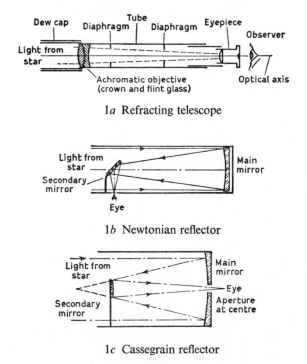

1a Refracting telescope

1b Newtonian reflector

1c Cassegrain reflector

Fig. 1. The telescope

cases we can reproduce in our own laboratories at least some of the characteristics shown by them.

For over a century now, physicists and chemists together with astronomers have been able to determine that the Sun is composed of the same elements as those which we find on Earth. This determination was made possible by the use of the spectroscope (fig. 2), the instrument which analyses light into its various wave-lengths. So a new branch of astronomy was born, the branch which we call astrophysics because it studies the physical conditions of celestial objects. Within the last century this new branch of astronomy has developed and expanded very rapidly and can now be called the 'new astronomy'. Astrophysics, thanks to the developments which in recent years have taken

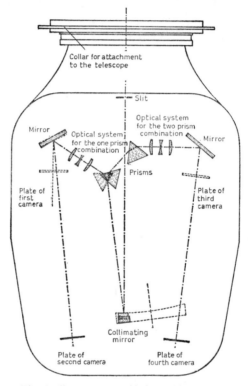

Fig. 2. Spectroscope (Asiago Observatory)

place in physics, is increasingly becoming an experimental science. Only a short step was necessary to make a fundamental discovery. Now we know that all the stars, and not only the Sun, are composed of almost all the chemical elements which man has discovered in the air, in water and in the terrestrial crust, although the physical conditions of stars are different from type to type. By means of this extremely powerful method of investigation man, therefore, has acquired the possibility of an-

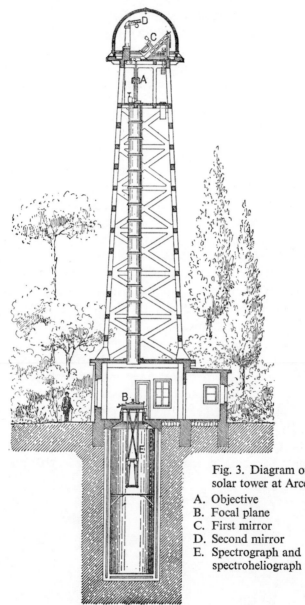

Fig. 3. Diagram of
solar tower at Arcetri

A. Objective
B. Focal plane
C. First mirror
D. Second mirror
E. Spectrograph and
   spectroheliograph

alysing in detail the celestial objects. He has also been able to study the physical characteristics of such objects and reach the very important conclusion that in the whole universe, or at least in that part of it which we can explore, there exists a unity of matter as there exists a unity of physical forces. This conclusion is not only important for science but for philosophy and religion as well.

One of the early pioneers of this branch of astronomy was Father Secchi, who wrote a book on 'the unity of the physical forces'. In this book he accurately predicted the wonderful developments of astrophysics which were to follow and sometimes even precede, the progress of physics.

Towards the end of the nineteenth century another pioneer, the great American astrophysicist, George Ellery Hale, founded the Yerkes Observatory of Chicago University. This observatory had what was at the time the largest telescope in the world, a refractor with a 40-inch objective. Hale soon discovered the necessity of using more powerful instruments and moreover he was quick to realize that such instruments had to be erected in selected places where the climate was more favourable than that of Chicago. Together with his associates he decided therefore to move to California, near the top of Mt. Wilson, and there he established a modern observatory with a large astrophysical laboratory. The instruments of this new observatory were soon to be followed by others. Hale first installed a horizontal telescope and two vertical telescopes (figs. 3 and 4), known as solar towers, exclusively for the study of the Sun. Later a reflector having a parabolic mirror of 60-inch diameter was followed by

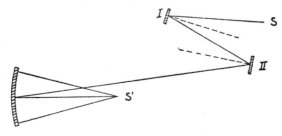

Fig. 4. Arrangement of the horizontal Snow telescope

22

the Hooker 100-inch reflector, which for many years was the largest telescope in the world.

Before long, the bright lights of Hollywood became a hindrance to the astronomers working with the 100-inch, and this, together with the need to study objects which were further away in the universe and hence very faint, led Hale to a grandiose project which was to be his last. Rockefeller and Carnegie joined forces, and under the auspices of the 'Caltech' (California Institute of Technology) a new observatory was erected on Mt. Palomar as a result of Hale's initiative. This observatory has the largest telescope in the world. It is named after Hale and it has an aperture of 200 inches (Plate 2).

We have recalled these facts because although so many American and European observatories have contributed so much to our increased knowledge of the universe, we must remember the inspired foresight of Hale who, in establishing these three great observatories, played such an important part in the development of astrophysics.

The Sun was one of the celestial objects which attracted a great deal of Hale's attention. The many discoveries made by him and by other astronomers about the Sun, showed the importance, indeed the necessity, of following the various phenomena it presents with the greatest possible continuity. In other words it became clear that it was of great importance to study the 'solar meteorology' as carefully as we do with meteorology on the Earth. In order to reach this goal it was necessary to establish international co-operation. Following Hale's initiative, this co-operation was at first established only in connection with solar observations, and gradually developed into the International Astronomical Union (I.A.U.). In recent years this co-operation has become so widespread that it has become possible to carry out such vast programmes as the International Geophysical Year (I.G.Y.) 1957–1958 and the International Quiet Sun Years (I.Q.S.Y.) 1964–1966. During these years, and not only in these years, the Sun has been under continuous observation by observatories distributed all over the Earth at various longitudes. The result of this is that the Sun is kept under observation 24 hours a day. The observations, taken with many instruments, may be visual, photographic or magnetic, and may also, as we shall see later on, involve radio telescopes.

Our knowledge of solar phenomena has increased considerably in recent years. We not only know their development but also the important influence that they exert on the Earth. There is no doubt that all this could not have been achieved without international co-operation.

It is a well-known fact that the Sun is a sphere of gas having a high surface temperature of about 6,000°C. The Sun is periodically subject to storms of a varying degree of intensity which manifest themselves as dark spots accompanied by spectacular eruptions, consisting mostly of incandescent hydrogen and other gases belonging to the elements of which the Sun is composed. Because of the turbulent constitution of the interior of the Sun, a seasonal cycle seems to exist, similar to what happens on the Earth, which has various seasons according to its relative position to the Sun during the course of the year. A 'season' on the Sun, which is due instead to internal causes, lasts on average for 11 years, so that in the course of these years we observe the maximum and minimum of the solar activity.

Since 1750, when fairly regular observations of the Sun began, 19 cycles have taken place. Some of these are longer and others shorter than 11 years, and in addition some of these cycles are more intense than others. The intensity of the cycle is measured by the frequency and violence of the phenomena which occur during the cycle itself and can be assessed by the number of spots which appear on the surface of the Sun. During the 19 cycles the highest maxima occurred in 1778 and in 1947. The Sun went through a minimum of activity in 1954, then gradually increased its activity and in 1958 reached a maximum which is the highest so far recorded since 1750.

Can we conclude from this that the Sun is moving towards some new manifestations compared with the past centuries? The previous observations mentioned cover a period which indeed is very small when compared with the age of the Sun. In any case, the intensity of the various cycles shows fluctuations which indicate the probable existence of a cycle of a longer period, superimposed on the 11-years' cycle. If we dared to make a forecast, based on the study of the previous cycles, we would say that the future cycles will be of lower intensity. We shall have the opportunity to return to this subject later on.

For a long time now it has been known that the Earth behaves like a large magnet, with magnetic poles which do not coincide with the geographical poles. From this it would be natural to conclude that other celestial bodies that are rotating and revolving around a centre of gravity, could also have magnetic fields. The difficulty appeared to lie in the detection and measurement of them. Hale discovered that all sunspots have strong magnetic fields which are very variable, and which have well defined polarities in the two hemispheres. Later on he was also able to discover the existence of a general magnetic field of the Sun. One of the remarkable characteristics of the magnetic fields of sunspots is that they are several thousand times more intense than the general magnetic field of the Sun, which in turn is only a little more intense than the terrestrial magnetic field. It was natural that following this discovery, astronomers should have asked themselves whether the stars, which are so much larger and hotter than the Sun, did not also have magnetic fields. Since the stars can be studied only as points it was obvious that only general magnetic fields could be detected. Such magnetic fields have been discovered in stars endowed with high speed of rotation. This discovery was made by H. Babcock, a director of the Mt. Wilson and Mt. Palomar observatories, by using special devices in conjunction with the 100-inch and the 200-inch reflectors. Some of these magnetic stars have a characteristic of changing polarity in short intervals of time, that is to say a change over between north and south poles.

Following these investigations it appeared probable that magnetic fields are a general property of matter in rotation and that, therefore, the whole Galaxy which rotates around its centre, could be the seat of large magnetic fields. Fermi and his colleague Chandrasekhar predicted by means of involved theoretical studies the existence and also the intensity of such magnetic fields in the spiral arms of our own Galaxy. This would support Fermi's hypothesis that cosmic rays had their origin in the acceleration to which are subjected the elementary particles forming cosmic matter scattered in space. This is a new branch of research which has appeared in recent years. It is known by the name of 'magneto-hydrodynamics' and studies the characteristics and properties of matter when subjected to magnetic fields.

The large telescopes (Plate 2) which we have available nowadays, can probe the depths of the Galaxy, which is the name we give to that very large system of stars in which the Sun with its planetary system moves. The Galaxy has a centre, or nucleus, which is hidden from us by the dense clouds of stars which we can see on a summer evening in the direction of Ophiucus and Sagittarius. With the help of radioastronomy we have discovered that the shape of the Galaxy is not as it appears to us, that is simply as a very large ring (Milky Way) surrounding the solar system. Indeed, from the nucleus of the Galaxy emerge wide spiral arms which consist of stars and bright interstellar matter. In this matter there is also cosmic dust and gases which absorb light and hence appear to us as large dark patches. We can conclude, therefore, that the Galaxy has the shape of an enormous Catherine-wheel, with a nucleus from which emerge the spiral arms in which stars seem to be born. In the system as a whole, the stars are not all the same, but apart from having large differences in brightness and dimensions, they also differ in temperature. Thus in it we find blue, yellow and red stars with different distributions of the chemical elements of which they are composed. The red stars, that is those with a relatively low temperature of the order of 3,000 or 4,000 degrees, are found in the central region of the system, while the hotter, very bright blue stars are more common in the spiral arms. Probably these blue stars have been formed more recently and may have a shorter life. With this knowledge available, modern astronomy started to investigate the degree of importance of our Galaxy in the universe. In other words whether beyond our Galaxy there existed other celestial bodies, what their distances and dimensions were and whether they were in any way related to the Galaxy by gravity or any other physical laws. Already towards the middle of the nineteenth century William Parson, the third Earl of Rosse, had discovered with his telescope, which was then the largest in the world, a celestial Catherine-wheel situated near the 'tail' of the Great Bear. Lord Rosse was able to detect from his careful and assiduous observations of this rather faint object, a very delicate spiral structure of almost geometrical precision.

When we observe the sky in the direction of the Milky Way, it is natural that the stars and the masses of gases of which it is composed, prevent us from seeing celestial objects which may exist beyond it. If, however, we point our telescopes in a direction at right angles to the plane of the Milky Way, or to any part of the sky away from the Milky Way, we should be able to discover whether other systems do exist. This has been done with powerful modern instruments, and they showed us the existence of such celestial objects beyond our Galaxy and confirmed the early discovery made by Lord Rosse. With our new telescopes we can probe space beyond our Galaxy. The first objects we meet outside our own system are those which can just be seen with the unaided eye, such as for instance the Andromeda Nebula (Plate 3). When this is observed through a powerful telescope, it shows without any doubt a very clear spiral shape. The Andromeda Nebula was compared with our Galaxy and it was found that the two systems are very similar as far as their physical constitution is concerned. As for dimensions, this presented a rather difficult problem because in order to determine the size of the Andromeda Nebula, it was first necessary to determine its distance, and this was not an easy task since there is no direct means by which this information could be obtained. Recourse had to be made to indirect methods, which in time have undergone so many modifications that the early determination of distance had to be increased by a factor greater than 2. In spite of the fact that the distance of the Andromeda Nebula is not accurately known, nevertheless it is almost certain that its dimensions are comparable, if not equal, to those of our Galaxy. It is estimated that the Andromeda Nebula is two million light-years away from us. At this distance, other systems of similar form but of different dimensions are found and their shape can be examined very carefully. At greater distances, photographs taken by the large telescopes with long exposure reveal the presence of thousands of millions of such systems. They can be easily identified on the photographs because they do not appear as sharp dots like the stars but rather as fuzzy images (Plate 4), some elongated, some thicker at the centre and thinner at the ends, or even very rounded but still nebulous in character and very different from the stellar images.

From what we can tell from a study of these galaxies which are of rather large dimensions, their composition appears very similar to that of our Galaxy, since blue, yellow and red stars are detected intermingled with cosmic dust and gases which appear as dark or bright patches. When we compare the various forms of these systems called 'extragalactic nebulae' or more simply 'galaxies', we discover an interesting sequence of forms, which leads us to assume the possibility of an evolution of these galaxies in the course of time. In their simplest form the galaxies appear as a gaseous mass, almost spherical in shape, which cannot be resolved into stars. This mass is often elongated towards the ends and crossed by a dark band. From this shape we gradually pass to those galaxies showing a more or less pronounced spiral form which can be seen in the sky either edgeways or face-on according to the orientation of their main plane to our line of sight. So far, however, we have not been able to establish the course of evolution of these galaxies from their birth to their full development and decay.

The expansion of the universe is another of the great mysteries which presents a challenge to man. All the galaxies appear to recede away from us with velocities which increase with distance, so that the furthermost galaxies appear to recede with the highest velocity. The faintest galaxies, namely those which are the furthest away from us which can be observed with the 200-inch reflector, are estimated to be at a distance of 2,000 million light-years from us. They appear to recede with a velocity of 37,200 miles per second, that is to say one-fifth of the speed of light. The simple conclusion follows that even if we had available such a powerful telescope which could detect galaxies still further away, those which have a velocity equal to the speed of light, would never be visible. This experimental fact, which does not seem to allow any other explanation, leads theoretical astronomers to discuss the cosmological problems of the structure of the universe. Do all these systems tend to run to a definite limit, or do they move towards an infinity which the human mind cannot grasp? We are not in a position at present to answer this fundamental question and perhaps we shall never be able to do so. Man, however, does not surrender easily and tries other ways to increase his knowledge and lift the veil covering this mystery. The means that man is developing are the artificial satellites and space probes, the electron telescope and radioastronomy. The electron telescope, which is still at a very early stage of development, employs electron optics and photomultipliers, which give a considerable amplification to the light received by the telescope. It is calculated that such a device added to the 200-inch reflector would give an increase in power of the order of 100 times. This will enable astronomers to obtain photographs of the remotest galaxies, which so far are beyond the present power of the 200-inch reflector.

In only a few years radioastronomy has made fantastic progress and now reveals a sky which so far has been completely unknown to us.

Already in 1916 Marconi in a paper to the Accademia dei Lincei, suggested that the natural electromagnetic waves, 'static' or 'atmospherics', were probably not all produced in our own atmosphere, and therefore the second name was not really accurate. Furthermore he stated: '. . . there is still some doubt whether these radio waves are produced by electric discharges in distant parts of the Earth, or by electrical disturbances in the interior of the Earth, or even by disturbances which originate outside the Earth and its atmosphere. Observations made simultaneously and for many years at radio stations in Europe and in America, have shown that in many cases the disturbances occur at the same time, I should say at the same instant, and with the same intensity both in America and in Europe. This would suggest that the point of origin of these radio waves must be at a very great distance, compared with the distance of about 4,000 km. which separates the transatlantic transmitters from us.'

In 1931 Jansky, a radio engineer working in the laboratories of the Bell Telephone Company, was investigating the effect of atmospherics on commerical transmissions on short waves. In the systematic records of these disturbances he noted that some of them followed the course of the Sun from sunrise to sunset. It was obvious that this was a case of a regular diurnal phenomenon, similar to the diurnal variation of temperature and barometric pressure for instance. Since the disturbance followed the course of the Sun, it was natural to think that the Sun was responsible for the effect.

Jansky continued the study of this phenomenon and soon discovered that the unknown source of 'noise' appeared to move towards the east and occurred four minutes earlier every day, so that after a month the 'noise' preceded the transit of the Sun at the meridian by almost two hours. It was thought, therefore, that not the Sun, but rather a celestial source further away, was responsible for the 'noise', and that its position happened to coincide with the position of the Sun when the observations were first made. The celestial co-ordinates of the source of 'noise' were the same as those of the star clouds in Sagittarius where the centre of the Galaxy is situated. Two years later the *New York Times* announced in large headlines the discovery of the emission of 'radio waves from the centre of the Galaxy' and soon after the B.B.C., linked to Jansky's receiver, broadcast this 'galactic noise' which sounded very much like the hissing noise of steam.

Radioastronomy was therefore born at the time of Jansky's early experiments, but for the first years following those experiments, its development was slow and uncertain. During the summer of 1942 in the midst of World War II, the radar installations, on the British coasts, which were in action to detect the approach of enemy aircraft, suddenly ceased to work normally and recorded a continuous 'noise' of a mysterious nature.

At first it was thought that this might well be an 'anti-radar' device of the enemy, but by tracking the actual position of origin in the sky, it was found to coincide with the position of the Sun. It soon appeared that the Sun was responsible for the emission of this 'noise'. Investigations showed that the radiation from the Sun consisted of a permanent or background emission with occasional bursts. By using receivers which were much more sensitive than those used by Jansky, it was found that the Sun emitted radio waves at various wavelengths. Since then the Sun has been under continuous observation, not only by means of optical instruments, as we have already mentioned, but also by means of radio telescopes. Gradually it was possible to establish the fact that the Sun emits radio waves of various frequencies in what is termed the 'radio spectrum'. These emissions are closely linked to the phenomena which are visible on the surface of the Sun, such as sunspots, eruptions of hydrogen and other elements, all of which can be observed with optical instruments such as solar towers (Plate 5).

The radio telescope has the advantage that it can register radio waves continuously, even when the sky is cloudy, and so it became possible to discover that these radio emissions are of various types. When the Sun is very active, great radio storms are recorded. These may last several hours and even days and seem to coincide with the appearance of large sunspots on the surface of the Sun.

In terms of temperature it is calculated that the intensity of radio storms is of an order ranging from 100 million degrees to 10,000 million degrees. Occasionally 'bursts' and 'outbursts' are recorded. These appear simultaneously with the solar 'flares', which are explosions of hydrogen, visible on the surface of the Sun as a very great increase of brightness in very limited regions, generally in the neighbourhood of sunspots. The fact that outbursts generally are recorded a little later than the appearance of the flares, is interpreted as being due to the time taken by the disturbance to rise from the lower layers of the solar atmosphere to the level of the corona. The velocity of ascent is of the order of 600 miles per second, which is approximately the velocity of the corpuscular swarms which reach the Earth from the Sun, and produce the auroral displays and the magnetic storms. The Sun appears to us visually as a disc with a well-defined limb and with luminosity which decreases from the centre to the limb. The radio Sun, however, seems to have much greater dimensions.

Radio emissions have also been detected from the planets of the solar system. Venus, Mars and Jupiter emit radio waves in the centimetre wave-band. At present very large radio telescopes are being built in order to study the radio emissions which reach us from the various regions of the Galaxy or from the galaxies. The largest radio telescope in use is that at Jodrell Bank (Manchester University) (Plate 6), which has also given valuable assistance in the tracking of artificial satellites. The parabolic reflector, or 'dish', has a diameter of 250 feet and has a depth of 60 feet at the centre where there is situated an aerial which in some ways is similar to those used for television.

The most important fact which has emerged from the radio observations of recent years, is that with very few exceptions, radio sources do not coincide with visible celestial objects. Therefore the name 'radio stars' which originally was used, has now been discarded in favour of 'radio sources'. These can be divided into two types. First we have sources of great intensity and which probably have diameters of the order of half a degree, namely comparable to the apparent diameters of the Sun and of the Moon. These sources seem to be located in the neighbourhood of the galactic plane. Radio sources of the other type are less intense, have angular diameters which are small and are distributed more or less uniformly all over the sky. Probably the radio sources of the first type belong to the Galaxy, while the others are beyond our galactic system. Great efforts have been made to try to find visual objects in the positions where the radio sources have been detected and the 200-inch reflector has been able to show that in some cases they coincide with some of the greater galaxies. In other cases it seems that the radio emission may originate in two galaxies which appear to be in contact with each other. Perhaps it is a question of two or more galaxies which are undergoing violent cataclisms in which

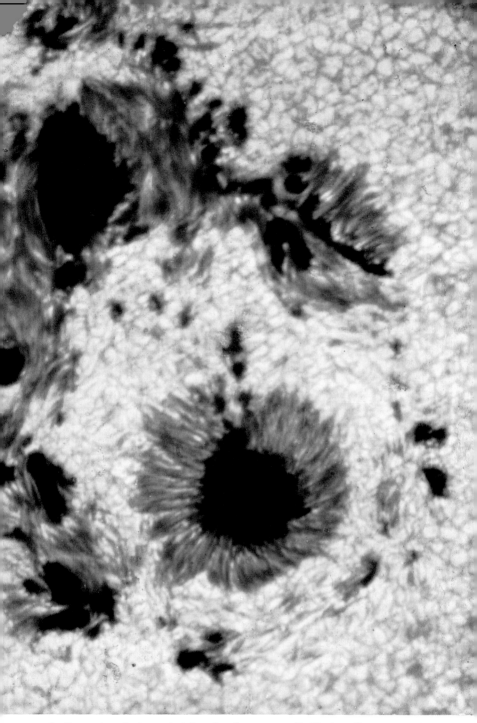

1. Sunspots: photograph taken by Stratoscope I. (Project Stratoscope, Princeton University, sponsored by NSF, ONR and NASA)

2. The 200-inch Hale telescope Mt. Palomar showing (*above*) observer in prime focus cage, and reflecting surface of 200-inch mirror; (*right*) the telescope pointing to zenith; seen from the south

3. Great Nebula in Andromeda M.31–NGC 224. Satellite nebulae NGC 205 and 221 also shown. 48-inch Schmidt Mt. Palomar

4. Cluster of galaxies in Coma Berenices. 200-inch Mt. Palomar

5a.
Exterior of the horizontal Snow telescope, together with the two solar towers at Mt. Wilson, California

5b.
The solar tower at Arcetri Observatory, Florence

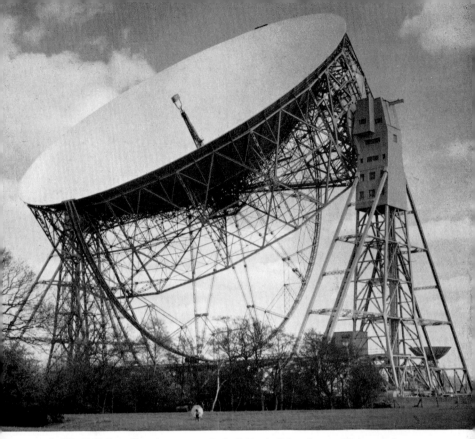

6. The 250-ft radio telescope at Jodrell Bank (Manchester University)

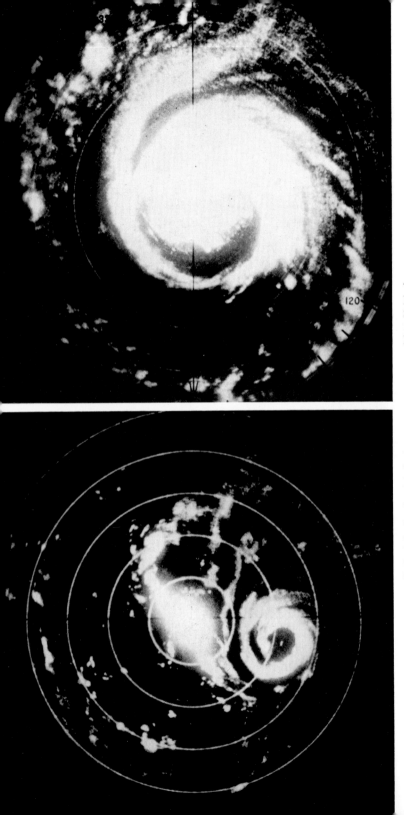

7a.
Hurricane of
September 21s
1948 at 11 h 31
E.S.T.

7b.
The same
hurricane at 1
10 min E.S.T.
The eye of the
cyclone appea
as a circular
vacant spot
surrounded by
rain-cloud ech

velocities of several hundred miles per second are developed in an interval of time of several millions of years.

Radioastronomy therefore opens a very important new field of investigation in modern astronomy. This is not only because we are discovering new events in the sky, but also because it is believed that the radio waves which reach us from the depth of space originate from distances much greater than those which can be studied optically by the 200-inch reflector, and which are probably of the order of a few thousand million light-years.

2 · Artificial Satellites, the Observatories in Space

It is certain that extremely useful and important data concerning our terrestrial atmosphere have been obtained by means of artificial satellites, setting aside those missiles with military applications and those which form part of the ambitious projects to reach the Moon or some of the near planets.

For centuries man has aspired to explore fully his planet and the space around it. Having almost completed the exploration of the Earth, he is now turning to the exploration of space, by using equipment and means which are becoming increasingly more complex.

On the one hand, the United States with their National Aeronautics and Space Administration (N.A.S.A.), and on the other hand, the U.S.S.R. with their Academy of Sciences are contributing gigantic effort and enormous cost towards the solution of this problem. N.A.S.A. was founded in 1958 with the specific task of directing all its efforts towards the exploration of space for scientific and peaceful purposes and for the benefit of mankind. The Goddard Space Flight Center was named after Robert Goddard (1882-1945) who in 1919 submitted a paper to the Smithsonian Institute of Washington, on a method to be adopted to reach great heights by using solid and liquid fuels and multistage rockets which, unfortunately, were soon adapted for military use, as has often happened with other inventions.

The Goddard Center, which is in Maryland, is mainly interested in the development of unmanned satellites for the exploration of space between the Earth and the Moon. Within this region of space many are the problems to be investigated, both in pure and applied science, and already we can safely say that a great increase in the knowledge of the space immediately around us has been achieved by means of the many satellites which have been launched. At the Goddard Center there are 2,700 scientists, including engineers, physicists, astronomers, geophysicists, mathematicians, geologists and radio engineers, who devote their time to the study of the universe.

Among the most important programmes of research we can mention three. The first concerns the study of the Sun, which so far we have been able to study only from the surface of the Earth. We have already had occasion to mention that one of the reasons for going high up in, or even above, our atmosphere is to avoid its absorption effect, on account of which a high percentage of the radiation emitted by the Sun is lost to us. The whole of the part of the radiation thus absorbed remained unknown to us until a few years ago. In recent years, however, we have been able to extend the investigations of solar radiation to extremely short wavelengths, such as the X rays and the Gamma rays. The great variety of phenomena which the surface of the Sun presents, such as sunspots, flares and prominences, studied from outside the terrestrial atmosphere, has already supplied us with further important knowledge about the star nearest to us. These investigations are very important not only because they enable us to understand better the constitution of the Sun, but also because of the great influence that the solar phenomena have on our atmosphere and indeed, on the Earth itself.

The second programme concerns the exploration of the interplanetary space which is dominated by the Sun and is not influenced by the Earth. In this region of space we can study the electromagnetic radiations emitted by the Sun, before they reach near enough to the Earth to be affected both by the terrestrial atmosphere and by the magnetic field.

The third programme refers to the exploration of the region nearer to the surface of the Earth and known as the 'magnetosphere'. We already know that on the Earth there exists a magnetic field which is regularly studied and measured by means of various instruments. Its major influence is felt on the surface of the Earth and in particular in the magnetosphere. It acts as a protective envelope which prevents the solar radiation from reaching the Earth in its original form. At the equator the magnetosphere extends up to about 25,000 miles, while at the magnetic poles the envelope is much thinner. Evidence of this is found in the formation of the aurorae which are produced by the Sun when it is very active, and which occur in the lower parts of the magnetosphere.

At the Goddard Center and at the other laboratories affiliated to it, various types of satellites are designed and built. They carry a variety of equipment according to the purpose each satellite is to fulfil. Generally speaking, the development of this equipment takes place at the various universities and centres of research which specialize in the various problems.

Thus satellites are built for specific observations of the Sun and of its radiations, and they are known under the name of O.S.O. (Orbiting Solar Observatory). For telescopic observations of stars in the ultraviolet region of the spectrum there are satellites which are 'astronomical observatories'. Satellites for other purposes have also been designed and launched. We can mention among these: satellites for the study of the composition, density, barometric pressure and temperature of the upper atmosphere; satellites for geophysical investigations and covering the study of the Van Allen belts, ionospheric phenomena and magnetic fields; satellites for the investigation of the concentration of electrons and radio waves propagation. All these satellites carry radio equipment in order to receive orders from the Earth and to transmit back the information they have collected in their journey through space. The energy necessary to maintain these instruments is supplied by means of solar batteries, which generally are in the shape of big wings which open up once the satellite is in orbit.

The orbits of the artificial satellites vary in size and shape from almost circular to very elongated ellipses and they may be equatorial or polar according to the type of investigation a satellite is to carry out. The American launching stations are situated on the Atlantic and Pacific coasts. Once the satellites are launched they have to be tracked during their journeys through space. For this purpose there is a network of about 20 stations scattered over three continents and three oceans and which are linked together. These stations follow the satellites by means of radar. They calculate the satellites' future positions, record the signals received which are the data of the observations made by the satellites and then transfer them on to magnetic tapes which in due course are decoded.

The meteorological satellites called Tiros and Nimbus, have already been of great help in observing the meteorological conditions over the whole of the Earth. They have been able to observe the distribution and amount of clouds, the position of storms and hurricanes which are in formation and then follow their paths (Plate 7). There is little doubt that the information obtained by these satellites over a long period of time, will enable meteorologists to understand better the mechanism of the weather and will lead, in due course, to long-term weather forecasts.

Among the satellites which are used for intercontinental communications we can mention Echo, Relay, Syncom, Telstar and Early Bird. The Echo satellites are very large balloons which reflect the radio waves emitted by some transmitting station on the Earth. The others carry electronic equipment so that the signals received from the Earth are first amplified and then sent back to the Earth. The results obtained with these satellites are already well known, as most of us have seen the long-distance television transmissions they have made possible.

Part I   THE SUN

1 · The Study of the Sun's Appearance

We can never look at the Sun directly when it is high in a clear sky but only through clouds or when it is low down on the horizon and it appears as a red or a reddish-yellow disc, sometimes without any markings and sometimes with some darkish spots. No one has ever doubted the fact that the Sun is a celestial body which emits the light and heat necessary to maintain life on Earth, even if in ancient times it was thought to be a satellite of the Earth, rather than the centre and ruler of the solar system.

Throughout the centuries, telescopes and many other instruments invented by man have enabled him to learn more about the phenomena which take place on the Sun's surface. At times people wonder how it has been possible, in spite of the distance of about 93 million miles which separates us from the Sun, to discover what are the elements which exist on the Sun and how its inexhaustible energy is produced. In asserting this knowledge it sounds as if the Sun were within our reach to enable us to study it in detail. In truth the Sun still presents many mysteries, some of which probably will one day be solved, while others will remain among those questions that man will never be able to answer.

The introduction of spectroscopic methods in the study of the Sun, namely the splitting of the Sun's light into its elementary colours, which began nearly a century ago, led to considerable progress in the knowledge of the Sun. This knowledge has

41

made even greater progress in the last 20 years or so, since the French astrophysicist Lyot invented the interference filter, and since cine technique was adapted to the study of solar phenomena. How the interference filter works is rather too complicated to explain here, but in order to understand its importance it will suffice to describe what we can see when we observe the Sun through it. When the interference filter is added to a telescope, the device allows through only a very narrow region of the spectrum of the Sun's light, let us say, for instance, in the red region of the spectrum. Now exactly in this region we can observe the distribution of hydrogen which is so abundant on the Sun, and see how this distribution varies with time. We can compare this study with that which an observer could make flying at a very high altitude above the surface of the Earth, observing the form, distribution and motion of the clouds above which he is flying (Plate 8).

The spectacle which can be seen through the Lyot filter, particularly when the Sun is very active, is really fascinating. On the disc of the Sun, which appears very red, we detect dark sunspots which are single or in groups, with their dark nuclei surrounded by a lighter penumbra and often by a bright ring. Among the spots we can detect irregular areas of various degrees of brightness which we call hydrogen 'flocculi'. These stand out against a background which can be compared to the ground covered with leaves. The flocculi are distributed in a fairly regular manner over the whole of the disc, but they assume characteristic shapes around the spots, which have the appearance of well-developed vortices similar to terrestrial cyclones. The flocculi have a right-handed or left-handed motion, namely clockwise or anti-clockwise. Very bright points suddenly appear generally on the edge of the spots or in the interior of the groups and between one spot and another. These bright points sometimes disappear after a few minutes, at other times they extend into larger areas or in the shape of thin filaments which may last several hours. These are the 'flares', points or areas of emission of intense ultraviolet radiation. The rapidity and the violence of these disturbances are an indication of the great amount of energy which develops

42

from these flares which occur occasionally, and sometimes with considerable frequency, in these disturbed centres.

Scattered here and there on the disc, without any apparent link with the spots, there appear curving lines of various length, which are always very narrow when they are in the central regions of the disc. These are the 'filaments' which reveal their nature if we follow them for a few days while they approach the edge of the disc. They appear to widen gradually and when they reach the limb of the Sun they seem to rise above the surface of the Sun in the shape of flames having varying degrees of height and width. These are the 'prominences' which at first were only observed by means of the spectroscope but now, with the help of the Lyot filter, can be observed in a much easier manner in all their extension and their splendour. These flames have shapes similar to trees or to the typical mushroom cloud which follows an atomic explosion. For some unknown reason they are narrow and long, and if projected on the disc of the Sun have the general appearance of filaments.

If we observe these prominences for a time, we detect that generally the brightest are very active and change shape rapidly if they do not disappear all together. Others, on the other hand, remain unaltered for days on end. Because of the rotation of the Sun we see them setting at the west limb of the Sun and sometimes we see them rise again at the east limb with a shape that is almost unchanged. Often large masses of hydrogen gas detach themselves completely from the limb and seem to hover above the Sun, like our clouds, but at heights of some thousands of miles they disappear slowly and fall back onto the Sun. If instead of the eye we use a cine camera for these observations, we can record on film the development of these phenomena which are sometimes very fast and at other times relatively slow. The actual development of the phenomena is generally accelerated when the film is projected so that we can have an easier understanding of the rapid and chaotic movement of these very large masses of gas. These movements, because of the nature of the atoms which are responsible for them, cannot be compared to volcanic eruptions on Earth, but are more like the electromagnetic phenomena which take place in our atmosphere during storms. From the cine records we learn that the gas appears to be projected from the interior of the Sun with great speed and violence to a great height above its surface, only to fall back upon it along predetermined trajectories on account of the magnetic fields existing on the Sun.

In addition to providing a view of the Sun's surface in red light, the Lyot filter can also be constructed to enable us to study a very narrow region in the green of the spectrum. These filters can be used for the investigation of another class of phenomena which take place in the upper solar atmosphere, namely the corona. Here, too, condensations of gases, not of hydrogen in this case, but of metallic elements, are formed very rapidly and pulsate and then dissolve. At present the relation between these phenomena and sunspots and prominences is not very clear.

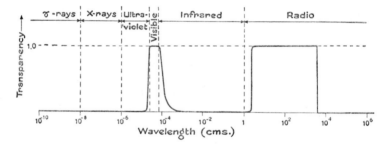

Fig. 5. Transparency of the Earth's atmosphere to electromagnetic radiations of different wavelengths

By means of photographic emulsions, which are more sensitive to the ultraviolet region of the spectrum than the human eye, and with the help of a filter suitable for this region, we can obtain an extremely interesting view of the Sun. These photographs show us the distribution on the Sun of the gases belonging to calcium, which in a solid form is an element extremely common on Earth. The appearance and distribution of this gas are very different from those of hydrogen. The solar disc in the regions which are not disturbed, appears covered by small grains which are alternately dark and bright and which probably are columns of gas ascending and descending on the surface of the Sun. Generally around the sunspots there appear irregular areas which are bright. These are the 'calcium faculae' which may sometimes be very intense and in which points or areas of great brightness tend to develop. These too are flares which may or may not coincide with the hydrogen flares.

In recent years radio observations have been added to the visual and photographic observations of the Sun and these reveal a new aspect and new phenomena. The Sun, which is a source of tremendous energy, emits radio waves of certain wavelengths (fig. 5). These emissions become much more intense when the Sun is active, so that if we connect a loudspeaker to a receiver we obtain a very loud noise accompanied by whistles of short duration, which often appear at the time when flares are observed optically. Observations of this type have only recently begun and have revealed to us a Sun which is very different from the Sun we are accustomed to observe visually or photographically.

2 · Filaments and Prominences on the Sun

In recent years the interest in and the importance of the study of what we could call solar meteorology have greatly increased. As in the case of terrestrial meteorology, solar meteorology follows and explains the mighty phenomena which disturb the surface of the Sun.

On Earth we have the seasons with various amounts of clouds, with rain, wind, storms and hurricanes. Similarly, disturbances occur periodically in the solar atmosphere, and modern instruments enable these disturbances to be observed and photographed. This atmosphere for us is transparent only for a very small part of the whole of its thickness and because of this our knowledge of what occurs in the solar atmosphere is very limited.

With the improvement of available instruments and the development of new ones, we have been able to observe many more phenomena than the sunspots which occasionally could be seen by the naked eye. In many cases the names given to these phenomena have been different according to the nationality of the scientist who first discovered them. As a result, occasionally, some confusion has arisen in the interpretation of these phenomena. In the end the names have been chosen which best correspond to the nature of the phenomena and these are translated in the various languages by the most suitable words.

Soon after the early observations of sunspots made by Galileo, by Fabricius and by Father Scheiner, Galileo discovered, near the solar limb, the existence of small bright patches which he called 'faculae'. This is a term in general use at present.

Chronologically there followed the discovery of the 'prominences', a word which expresses very clearly the nature of this particular phenomenon. When the spectroheliograph was invented in 1892 by Hale, and almost at the same time by Deslandres, the photographs obtained with it in the monochromatic light of ionized calcium or of hydrogen, showed the existence of something similar to the faculae.

Bright clouds of ionized calcium or of hydrogen are formed in that part of the atmosphere of the Sun which we call the chromosphere, just as in our own atmosphere clouds of water vapour are formed. Our own clouds often appear bright because they are illuminated by the light of the Sun. In the case of the Sun, its clouds appear brighter than the background of the disc on which they are projected. This is due to the higher degree of excitation of the gas which is present in the disturbed regions of the Sun, where they appear, and to sub-chromospheric or sub-photospheric phenomena which we cannot observe directly. To this phenomenon Hale gave the name of 'bright flocculi', while Deslandres gave the French name of 'plages faculaires'. In this particular case we must say that the term chosen by Hale is much better than that chosen by Deslandres, because it avoids the confusion with the faculae. It is true that the latter are often present in the same disturbed regions, but they differ from the bright flocculi in size, in intensity and are observed by a different method.

Another phenomenon was also discovered both by Hale and Deslandres in the monochromatic photographs taken of the Sun. Hale called it 'dark flocculi' while Deslandres used the French word 'filaments'. Both these words refer to some of the characteristics of the phenomenon. The English word refers to the intensity of the flocculi which is less than that of the general background of the solar disc. As we have already mentioned, these markings appear like narrow, long dark clouds, like filaments projected on the disc of the Sun. Here we could draw some remote analogy with the dark clouds, sometimes very dark clouds, of the fractonimbus type which are visible in our terrestrial atmosphere at the time of heavy storms. Although we know that the phenomenon is very different in the case of the Sun, nevertheless in both cases the effect is due to absorption of the radiation received by the gas forming the cloud. In the case of the Earth, the radiation absorbed is the light of the Sun. In the case of the Sun, the radiation absorbed is that emitted at given wavelengths by the atoms of hydrogen and of calcium. Since the dark flocculi have almost always the shape of filaments, but characteristics and motion very different from the bright flocculi, it is more convenient in this case, to adopt the name given to them by Deslandres, namely filaments, rather than use Hale's term (Plate 9).

The real nature of the filaments, which was soon suspected by many observers, was definitely confirmed by Ellerman, one of Hale's associates. His work was based on the remarkable spectroheliograms obtained by him, nearly 50 years ago, with the Snow spectroheliograph at Mt. Wilson. At that time, cinematography had not yet been adopted for solar observations. Today the method is very common and enables us to follow the development of solar phenomena in great detail. The spectroheliograms obtained by Ellerman were taken at regular intervals of time. Thus the filaments were recorded when they appeared at the eastern limb of the Sun and followed across the face of the disc of the Sun until they disappeared at the western limb on account of the solar rotation. These spectroheliograms gave enough evidence to clarify both the nature and the shape of the filaments. At first, at the limb, it is clear that the filaments are simply prominences of various degrees of height and extent. By means of the spectroheliograph it is possible to study the gases which are most abundant in the filaments, namely hydrogen and calcium; the hydrogen is observed by the separating and studying of the red line known as H_α and the calcium by studying one of the separated lines of ionized calcium known as K_{23}. These gases are ejected from the Sun and often, when they reach the high level of the prominences, they are so rarefied that the photosphere underneath can be seen through them (Plate 10).

The bright flocculi, which are brighter than the chromosphere, also rise sometimes in shapes similar to the prominences, but generally they are not as high as the filaments. It is therefore better to use the name of bright flocculi for this phenomenon, which indicates their characteristic of being similar to clouds or to irregular bright patches. On the other hand, the name of filament for the dark flocculi indicates clearly their characteristically elongated form.

We can talk of filaments which are seen projected as such on the disc of the Sun and which become prominences at the limb. Because of the various positions in which we can see them, we can determine their shape, their dimensions and their height above the chromosphere. Statistical studies of the form, frequency, distribution and motion of both filaments and prominences, have been carried out by many astronomers. As far as prominences are concerned, we have available observations going back to 1869, that is for about nine solar cycles, but in the case of filaments, the observations available are only limited to no more than six cycles.

The typical shape of filaments in space is that of a row of trees which rise above the chromosphere with wide foliage and branches which join together to form arches between each tree. The shape and appearance of the filaments can also be compared to a horse's mane, as was suggested by Lyot.

When a filament is formed on the Sun, the end which is nearer to the solar equator is generally at the same latitude as that of the sunspots and is often associated with new groups of sunspots. This may indicate the existence of a single centre of activity, which lies below the photosphere and comes up to the surface and manifests itself in the form of sunspots, flocculi and filaments. The filaments in their first stage of development are generally small and winding and are subjected to violent internal motions. Often we can notice that some material is ejected from the highest parts of the filaments. The centres of attraction which are near or within the surrounding sunspots, suck back this material which follows long, curved trajectories.

If the filaments last for several days, they appear to move

slowly towards the poles and at the same time become distorted and more curved. They reach their maximum development at their fourth transit on the visible hemisphere and that is approximately in three months. Their curvature is due to the well-known fact that the period of rotation of the Sun increases as the latitude increases. The filaments, as they move towards the poles, begin to be subject to this retardation. They become elongated and they stretch along the parallels. It is this peculiar orientation which explains the great variety of shapes assumed by the quiescent prominences. An equatorial filament, for instance, which generally lies in the direction of a meridian, is seen face-on when it reaches the limb. During its progress from the east limb to the central meridian and from this to the west limb, we can determine from these various positions, both the real form in space and the dimensions of the filament. On the other hand, a filament which is situated at a high altitude, lying in an east–west direction, will be seen edgeways at the limb, or almost in the shape of a pyramid or of a single tree. At the time of the fifth rotation of the Sun, the filament generally disintegrates and its equatorial end is the first part to disappear.

Many filaments or prominences undergo violent interruptions in the course of their ordinary development. Often a dense filament disappears, only occasionally, to reappear, after many days, in its original form and density. Sometimes we witness a real and proper explosion and the filament is subject to violent internal motions either in its totality or over the greater part of it. It rises above the chromosphere and, after a few hours, during which it moves with a constant ascending velocity, it undergoes sudden impulses until it leaves the Sun altogether. On other occasions the filament may follow a curved trajectory, and then it is swallowed up by the solar surface as if it were subjected to the influence of a very intense magnetic field.

The length of life of the filaments, their reappearance in the same position on the Sun, together with their migration towards the poles of the Sun, indicate that the gases of which the filaments are composed, come from some disturbance of long duration which has its origin in sub-photospheric levels. If this hypothesis is accepted, we can think of the filament-prominence as travelling in the upper chromosphere in such a manner that it is possible to observe it. The cause of this motion could well be a disturbance at a lower level. Often the filaments are transparent and in fact the light of the chromosphere below can be seen when the filament overtakes the solar limb and is seen as a prominence. Moreover, at the minimum of the solar cycle, prominences often are not seen as filaments on the disc.

From the spectroscopic images of the solar limb published since 1869 in the *Memorie degli spettroscopisti italiani*, and from other similar observations, several investigators, among whom Riccò of Catania in particular, have discovered that prominences can be divided into two groups. The first containing high-latitude prominences and the second containing low-latitude prominences, the dividing line between these two groups being at a latitude between $\pm 30°$ and $\pm 40°$. This, as is well known, is the limit of latitude for sunspots. Observations do not show the existence of any structural difference between the two groups, nevertheless the division can be justified by the fact that prominences appear to start their migration from a common mean latitude. High-latitude prominences migrate towards the poles of the Sun, while the low-latitude prominences follow the migration of sunspots towards the equator, during the course of the 11-years' cycle of solar activity.

Since filaments are nothing more than prominences, they naturally follow the same laws. Statistical investigations, like those carried out in recent years at Arcetri, explain some of the facts which were already known from the study of the prominences. There is some evidence that the low-latitude prominences may themselves be the origin of the high-latitude prominences. L. d'Azambuja has made an important study of the spectroheliograms obtained in the lines of calcium K_2 and K_3 and of the *Cartes synoptiques de la chromosphère solaire* published by the Meudon Observatory. He reaches the conclusion that the polar filament-prominences are probably the same filaments as the equatorial ones which migrate towards the poles of the Sun and reach high latitudes on account of a general circulation in the upper solar atmosphere.

From the observations of both filaments and prominences it can be deduced that at the beginning of the 11-years' cycle, that is at the time of minimum activity of the Sun, sub-photospheric disturbances begin around latitude $\pm 40°$. These disturbances show themselves on the surface of the Sun as pores or small sun-spots, which are an indication of the beginning of a new cycle. It is in this region of latitude that bright flocculi, filaments and prominences also appear. As the cycle progresses, the general disturbance migrates towards the lower latitudes, producing sunspots, flares, eruptive prominences and filaments, which all slowly reach the equatorial regions. On the other hand, the fila-ment-prominences migrate slowly towards the polar regions, which are reached at the time of the maximum of the 11-years' cycle.

Signs of the beginning of the new 11-years' cycle can appear from three to six years before the maximum, according to the greater or smaller intensity that the particular cycle will reach. It follows that after the maximum is reached, the activity of the equatorial region may last from eight to five years, while it is slowly approaching the equator. Accordingly, sooner or later, a new disturbed region will appear at latitudes $\pm 40°$ and a new cycle of solar activity will begin.

3 · Solar Flares

When considering solar flares we must remember that this phenomenon is not new, having been known already for nearly a century. What is new is the name, which has been introduced only in recent years, after the discovery that the appearance of a flare is followed almost simultaneously by a disturbance in our ionosphere and by remarkable effects on the terrestrial magnetic field.

Very intense flares have been observed only very rarely in white light and by direct observation. One of such cases was the phenomenon recorded in a large group of sunspots by Carrington and Hodgson on September 1st, 1859. They saw all of a sudden two very bright and small points appear, which had an intensity much greater than that of the photosphere nearby. They described it as a very extraordinary light which appeared to be in motion, which lasted five minutes and which disappeared as suddenly as it had appeared. This phenomenon was considered to be an exceptional eruption because of its very remarkable intensity, and also because similar observations are rare. Since Hale's invention of the spectroheliograph in 1892, it became possible to obtain monochromatic photographs of the Sun, mainly in hydrogen or calcium light, and many eruptions of the same type as that observed by Carrington and Hodgson have now been observed.

To obtain a better understanding of the nature of the phenomenon we had to wait until 1936, when the solar observations became much more continuous thanks to international co-operation. At present these eruptions are being observed in all their phases and their developments are carefully recorded, particularly by means of cine cameras. The nearest terrestrial analogy to these eruptions would be the extremely intense light released in our atmosphere by the explosion of a nuclear device or by natural electrical phenomena such as lightning during a thunderstorm. The latter are phenomena which take place at various heights above the lower layers of our atmosphere. Likewise the solar flares occur in the lower parts of the chromosphere, but above the photosphere where the umbra and penumbra of sunspots are formed.

Flares (Plate 9) generally appear within the bright flocculi of hydrogen or calcium which surround the sunspot groups, and often just on the edge of the umbra, where the penumbra begins. Flares shine for a few minutes and then disappear, only to reappear a short time later almost at the same point, or at least in the same region. The simplest form of a flare is that of a point of light, but sometimes, when the sunspot groups are very active, particularly during the maximum of the solar activity, the flare may acquire various forms and its area may be quite large. As an example we may mention the exceptional flare which occurred on the Sun on July 25th, 1946 at 17.32 U.T. and which was photographed in Hα light at the Meudon Observatory. The flare occupied the whole area between two sunspots which formed a bipolar group. The whole region was greatly disturbed, for the duration of nearly an hour, by an intense explosion of light produced by hydrogen. This explosion took the form of a vortex around the sunspots and must have reached a very high temperature following some phenomenon which we call 'magneto-hydrodynamic', and the nature of which is still not well understood. While small flares, point-like in appearance, have a diameter of the order of 6,000 miles, the more extensive ones may reach diameters a hundred times larger or even more. During the time of great solar activity, near the maximum of the 11-years' cycle, it is possible to detect on the visible hemisphere of the Sun as many as ten flares a day, of various intensities and dimensions. Point-like flares, when observed at the limb of the Sun, have the appearance of thin flames, other flares appear as very bright extensive regions, which may have a height of a few hundred miles above the mean level of the chromosphere.

We may ask ourselves what is the relation between flares and filaments. It is rather difficult to say. We know that filaments can be identified with the prominences at the limb of the Sun and that therefore they are a phenomenon occurring at a high level, thousands of miles above the chromosphere, and it may be that they are a consequence of flares, once the latter have disappeared. There is, however, one objection to this theory. It is known that filaments are formed and observed generally at higher latitudes than those where flares occur. In these regions, normally, we do not find sunspots which are the seat of flares. It has been suggested that when a flare is extinguished, and its life as we have seen is very short, hydrogen and the other gases of which the flares are composed, move towards higher latitudes where they can remain for a considerable time in the form of filaments. Because of the high altitude reached above the chromosphere by the gases, their temperature will be comparatively low and therefore they will absorb the light from the chromosphere and appear dark against the background of the Sun's disc.

When the Sun goes through a minimum of its activity, several months may pass without a single flare being observed on its surface. With the increase of solar activity which reached another maximum in the years 1957–1958, a very large disturbed area was observed in January and February 1956. In a region of the Sun centred around 20° of latitude north, and extending well over 60° in longitude, more than 70 sunspots of various sizes were observed, as well as calcium and hydrogen flocculi.

The phenomena which take place on the Sun are of increasing interest to man, not only for purely scientific reasons, but also for practical reasons. This explains why a co-ordinated effort is made internationally to keep a continuous watch on the Sun, from stations scattered all over the Earth. In addition to this network of observing stations, there has also developed in recent years a rapid system of exchange of information, drawings, photographs and the publication of regular bulletins giving a summary of the observations. This follows in some ways the example set by the international exchange of meteorological information concerning the weather all over the Earth and which enables meteorologists to produce their weather forecasts. Among the most important of the solar bulletins we may mention the *Zurich Quarterly Bulletin on Solar Activity* and the daily diagrams published by the Fraunhofer Institute of Freiburg which is a summary of all observations made in Italy, Switzerland, France, Germany, Spain, Turkey, India, Japan and Australia. In addition there are several other centres which publish similar information, among which we may mention the Arcetri Observatory.

4 · The Solar Wind

The American spacecraft Mariner II was used during its journey towards Venus to send back to Earth information about the conditions it met in the interplanetary space through which it travelled. One of the questions investigated by Mariner II was that of the 'solar wind', the existence of which was known from the phenomena which it produces on our own Earth.

The solar wind consists of swarms of charged particles, brought to a very high state of excitation, which in terms of temperature is estimated to be higher than 500,000°C. These swarms of particles are emitted continually by the Sun. They travel through space with a velocity ranging between 200 and 500 miles per second, and collide with the terrestrial atmosphere.

These swarms consist of ionized gases, called 'plasma' by the astrophysicists. The plasma is ejected into space by the chromosphere and by the inner corona of the Sun. Its velocity, density and temperature can be measured. The plasma consists of a mixture of electrons, helium nuclei and nuclei of heavy atoms which have lost several of the electrons surrounding them. Because of the speed with which this gas leaves the Sun, it has been called 'solar wind'. Its existence had already been detected on Earth because it is responsible for the magnetic storms that we record on Earth, the 'fade out' in radio communications and for the auroral displays.

Several artificial satellites had already recorded the presence of the solar wind in the neighbourhood of the Earth, but

Mariner II was able to show not only that the solar wind exists at a distance of several million miles from the Earth, but also to give us some information on its constitution. Special instruments aboard Mariner II relayed to Earth the velocity of the solar wind as well as the intensity of the energy with which its particles are endowed. Mariner II obtained a great number of observations from which we deduce that the energy of the particles of the solar wind is certainly much smaller than the energy of the particles of cosmic rays. In other words; cosmic rays consist mainly of protons, moving with a velocity of the order of light, that have a very penetrating power, while the penetrating power of the slower solar plasma is much less.

In the regions of the solar system the number of particles, namely the density of the solar wind, is much greater than that of the cosmic rays. It is for this reason that the solar wind produces phenomena which we observe on Earth, while we do not notice the presence of cosmic rays. When the Sun is not very active, the velocity of the solar wind is approximately 250 miles per second and its temperature is not higher than 2,000° or 3,000°C. On the other hand, when the Sun is very disturbed and active then clouds of plasma are emitted which are endowed with much greater velocity and temperature.

We have already discussed the solar phenomena which the astrophysicists call flares. They appear as chromospheric eruptions covering areas of various size and exploding in the regions occupied by sunspots. Their life may range from a few minutes to hours and the eruptions rise above the level of the chromosphere. Flares are subjected to violent and irregular motion but they are always affected by the presence of the magnetic fields which develop within the sunspots. Mariner II in its journey through interplanetary space, has been able to record the sudden arrival at great velocity of dense clouds of plasma and in some cases it was possible to establish a link between this phenomenon and the appearance of flares on the Sun. The correlation between flares and disturbances of the terrestrial magnetic field was already known and had been studied in great detail during the I.G.Y.

We now begin to understand a little better how the solar phenomena are produced and how they are linked with terrestrial manifestations which appear after a certain time lag. There is still, however, a great deal that we do not know or understand and it is certain that further launchings of space probes towards the planets or towards the Sun, will help us to achieve a better understanding of many questions which are at present still unanswered.

From the present state of our knowledge it seems certain that the solar wind has the power of suppressing magnetic fields which it meets in its path. As a result of this the terrestrial magnetic field is not uniformly distributed around the Earth. It becomes somewhat flattened on the side facing the Sun and is elongated on the opposite side, where in collecting particles which are captured by the magnetic field, it forms a tail similar to that of the comets.

5 · The Solar Eclipse of June 19th, 1936

It may seem an extraordinary thing that scientists should make many calculations, prepare complicated and expensive instruments and travel great distances in order to observe a celestial phenomenon which actually lasts only a few minutes. It is not, however, the duration of the phenomenon which determines its importance, but rather the results which can be obtained during that short interval of time, and only during that time, when everything has been properly organized and prepared. This justifies the astronomers' efforts to grasp every opportunity of observing such a rare event as a total eclipse of the Sun which can only be seen in limited regions of the Earth.

A confusion seems to exist in many people's minds between total and partial eclipses of the Sun and even lunar eclipses. It is perhaps necessary to stress here that only during a total eclipse of the Sun can the outer layers of the solar atmosphere be studied as they are only visible at this time. The region of the Earth where such a phenomenon can be observed is very limited. This is due to the nature of the phenomenon itself, that is the relative size and distance of both the Sun and the Moon from the Earth. On account of this, astronomers have to travel to places on the Earth from where the total eclipse is visible, and the location of these places can be accurately determined in advance by astronomical calculations. Once the region of the Earth is known from where the phenomenon is visible, it is natural that astronomers will choose a particular place within the region, where the meteorological conditions are most promising, namely where there is the greatest probability of clear skies, since even a few clouds would prevent the delicate observations. Another question which has to be taken into account in the choice of a suitable site for observations, is that the Sun should not be too low above the horizon at the time of the eclipse, as it is at sunrise and sunset.

The path of totality in the case of the eclipse which occurred in June 1936 was about 6,000 miles long and about 70 miles wide. It was of particular interest to the Russian astronomers since it crossed almost completely the vast territory of their country. In Greece the eclipse was visible at sunrise, while on the western side of the Black Sea it was visible at seven o'clock, with the Sun at an altitude of about 25°. Across the European part of Russia, the Urals and Siberia, the Sun was eclipsed increasingly later, with the maximum of totality of about 151 seconds over Lake Baikal, ending near sunset north of Japan on the island of Hokkaido. According to the information supplied by the Russian scientists, the greatest probability of a clear sky in the summer was to be expected in the Orenburg district near the Ural river.

Several expeditions from many countries had prepared very elaborate programmes of observations and various instruments for the occasion. Foremost among these were the Russian astronomers, who were able to organize within their own territory a very large network of stations stretching along the whole of the path of totality which crossed their country.

In Italy the tradition of solar research had started with Father Secchi who, during the total eclipse of the Sun in 1860, which was visible in Spain, was the first to obtain a photograph of the eclipse by means of his telescope having a focal length of approximately 80 inches. Together with De la Rue, he established some fundamental facts. Firstly, that the prominences, visible around the edge of the Sun during the totality, were real and not optical illusions as others thought, and that they belonged to the Sun. Secondly, that the corona was also real and more fully developed around 45° of latitude and around the equator than at the poles.

Father Secchi was less lucky in his expedition to observe the total eclipse of 1870, visible from Sicily, where in Augusta he had set up his station. The eclipse was to take place on December 22nd and from Augusta he wrote to a colleague in Rome: 'Do not forget me while I am here, for my sins, isolated from the rest of the universe, on this rock known as Augusta. Let us hope that everything will go if not well, at least not too badly.' On the actual day the Sun appeared, but so did the clouds, and the whole of the programme of observations was spoilt by clouds right from the beginning of the totality. Tacchini, Nobile and Lorenzoni had set up their station at Terranova and were luckier. They confirmed the existence of the corona and the presence of the reversing layer. The latter was seen also in the cloud gaps by Father Secchi towards the end of the totality. For the eclipses which followed we can remember the valuable work done by Riccò, Palazzo, Mengarini, Horn d'Arturo and Taffara in various expeditions in Spain, Russia and Somaliland. Their very successful observations made a considerable contribution towards the knowledge of both the chromosphere and the corona.

The typical problems of solar physics, which can be solved only during the short time when the light of the Sun is eclipsed by the lunar disc, are mainly those relating to the constitution, the intensity and the shape of the chromosphere and of the corona. By chromosphere we mean that upper part of the solar atmosphere in which remarkable disturbances of varying degrees of intensity can be seen, according to the state of activity of the Sun.

It is well known that when the disc of the Sun is observed by means of a spectroscope, the spectrum consists of a continuous band of colours from the violet to the red with dark lines, the Fraunhofer lines, superimposed upon it. When the disc of the Moon covers almost completely the solar disc, the latter appears to be surrounded by a thin layer of gases which we call the chromosphere. This observed with a spectroscope will show, with the decrease of the light of the continuous spectrum, almost the same Fraunhofer lines, but this time they will appear bright instead of dark. The study of these bright, or emission, lines as they are called, is of the greatest interest to the astrophysicist because it leads to the knowledge of the constitution, the temperature and the density of the chromosphere. Above this the corona extends to great heights and is visible only during the time of totality.

The beautiful corona which surrounds the Sun, and whose origin is still a mystery, changes its form during the course of the 11-years' cycle of the solar activity. Very little is known about its spectrum, which consists of a few bright lines. It is not surprising, therefore, that no opportunity is missed to photograph and study the corona with the hope of discovering which gases are present and what are their temperature and density at various distances from the Sun. In addition it is hoped that the investigations will give some indication of the mechanism which governs both the existence and the periodic changes of the corona. By means of photographs taken with suitable filters, the form of the corona and the intensity of its various parts can be obtained. In addition, when these photographs are taken from stations which are well apart within the path of totality, it is also possible to determine the changes which take place in the corona as a result of the disturbances produced by the prominences, as well as the motion of matter in the corona.

In order to investigate some of these problems, three special instruments were constructed at Arcetri for the Italian expedition which went to Russia in 1936. These instruments were a spectrograph, a spectrograph for the chromosphere and a coronograph.

The first instrument consisted of an equatorial mounting and drive to which were fixed two spectrographs, one on each side of the declination axis. One of the two spectrographs had optical parts made of special glass suitable for the ultraviolet end of the spectrum, while the other spectrograph was intended to be used for the visible region of the spectrum including the red region. Thus the combination of these two instruments enabled the simultaneous observation of the whole spectrum of the corona.

The optical parts of the spectrograph for the chromosphere

63

consisted of equipment already available at Arcetri, namely the two objectives of 10 and 11 inches aperture made by Amici in 1840 and the large grating which is part of the solar tower and was ruled by Jacomini in the workshop at Mt. Wilson Observatory. By means of a coelostat, the larger of the Amici objectives gave an image of the Sun, 2 inches in diameter, on the slit of the spectrograph. The light then passed through the second Amici objective and reached the grating which gave an extended spectrum of the Sun.

The third instrument was a coronograph, that is to say a photographic camera with a long focal length (approximately 100 inches), with two objectives in front of which were placed two filters, one violet and the other yellow-red. The images produced by this camera reproduced the appearance of the corona in the violet and red region of the spectrum respectively. The coronograph, placed horizontally, received the light of the Sun fed by a coelostat consisting of two plane mirrors. For the sake of brevity the three instruments were nicknamed 'giraffe', 'elephant' and 'gazelle' by the members of the expedition.

Of course these were not the only instruments required by the expedition. Several ancillary instruments were also available, such as photometric lamps, subsidiary small telescopes and batteries, which all together were required for the photometric calibration of the plates obtained during the eclipse. The calibration of the plates consisted either of images of a luminous source, or of a spectrum given by a lamp of which the intensity and the distribution of the energy in the various parts of the spectrum were known. By means of this calibration it was possible to determine the characteristics of the light which, originating from the chromosphere or the corona, was recorded by the photographic emulsion.

The expedition, consisting of Abetti and Righini of the Arcetri Observatory, Taffara of the Catania Observatory and Mrs. Abetti, left Italy on May 16th with all the scientific equipment, for Moscow. The Soviet Government, the Russian Academy of Sciences and Intourist had organized very efficiently all that was required for the transport of the equipment once in

Russia and the setting up of all the stations both national and foreign, along the path of totality.

In Moscow a special railway coach was available for the Italian expedition and for the Czech expedition which was to join us later on. The coach, provided with sleeping accommodation and cooking facilities, became our home, mobile first and stationary later, for 31 days, and was coupled to the train of the Siberia and Turkestan line, direct to Alma Ata. It reached Orenburg and from there it joined a secondary line, the Orenburg-Orsk and on May 24th reached the small station of Sara at about 30 miles from Orsk. Here, in order that our coach should not be in the way of the main traffic, it was shunted into a siding especially laid for the purpose. Near there, a wooden hut had been built, above which there was a tank of water, so that this hut was used as bath and shower by the members of the expedition. In addition we erected some tents which we had carried with us.

The small railway station of Sara was in the middle of the desolate steppe which surrounded it. The village of Sara consisted of a few mud huts, while a little further on, a very large granary and a Soviet farm formed two more inhabited centres. Professor Tikhov of the Pulkovo Observatory, in charge of the expedition from that observatory, was there to welcome us on our arrival. Another Russian station from the Pulkovo Observatory, manned by a larger number of astronomers, had set up their equipment at Ak Bulak, about 90 miles further west. Here there was also a large expedition from the Harvard Observatory (U.S.A.). Tikhov had set up his camp $3\frac{1}{2}$ miles from Sara station, in a small wood of birch trees which were the only trees visible for miles around. As far as we were concerned, we found it more convenient to set up our instruments close to our coach-home. Since our instruments had arrived punctually the day before, we immediately got down to the work of setting them up.

We realized at once that in the desolate steppe where a strong dusty wind is always blowing, our tents would not give adequate protection to the instruments. We therefore built a strong hut to house the three instruments with a partitioned part to be used as a dark room, while our tents were to be used for the stores. The extremes of temperature, more dangerous for our wooden instruments than for the personnel, could range from 100°F. in the shade during the day, to 32°F. during the night. Astronomical observations enabled us to orientate the three concrete pillars for the instruments as well as the wooden hut. This was completely open towards the part of the sky where the eclipse was to take place, and allowed us to observe the Sun for several hours during the day. The building of the hut was completed in a few days and that left us about twenty days before the eclipse took place. These days were not too many for all we had to do, that is erecting and checking the instruments, carrying out the necessary photographic tests in order to determine the exposure times and calibrating the photometric lamps.

The weather was generally very good until June 15th, but on June 16th and 17th it broke. Strong winds, accompanied by heavy rain, put to the test our temporary observatory and our tents. June 19th started as a beautiful day, but an hour before the eclipse was due to begin, the sky was quickly covered by clouds, so much so that our hopes of being able to observe the eclipse became very slender. We were however lucky. As the time of the eclipse approached (8.18 local time) the sky suddenly cleared completely, enabling us to obtain a very good observation of the phenomenon. Immediately after the eclipse, the sky clouded over again and in the afternoon we had a heavy thunderstorm with heavy rain.

The long and exhaustive preparations and the rehearsals carried out for six or seven consecutive hours, enabled us to fulfil the programme of observations we had set ourselves, during the approximate two minutes that the actual eclipse lasted. As soon as the Baily's beads appeared, one second before the beginning of the totality, due to the inequalities in the lunar surface which breaks the thin crescent of light into separate brilliant dots, the work began. I was able to see the Baily's beads very clearly on the slit of the ultraviolet spectrograph of the 'giraffe' by means of the small guiding telescope. At the same time Mrs. Abetti began the counting of the seconds with the help of a metronome and Righini started a series of exposures of his plates with the 'elephant' in order to obtain the flash spectrum, or the spectrum of the chromosphere. Immediately afterwards Taffara, who was manning the 'gazelle', began a series of exposures to obtain the photographs of both the inner and outer corona in violet and yellow-red light. Six or seven seconds after the beginning of the totality, I obtained two plates at the two spectrographs of the 'giraffe' by pointing the slits at the inner corona, as well as a photograph with a teleobjective in order to record the more external plumes of the corona. These three plates were exposed for a duration of 110 seconds out of the 118 seconds that the totality lasted. Righini and Taffara succeeded in obtaining 14 plates during the totality. Righini was helped by our interpreter Miss Vagner, and Taffara by Mr. Scoricov, the representative of Intourist, who had directed the installation of our camp.

Immediately after the eclipse and during all the following night, we took photographs with the three instruments. Photographs of the spectrum of the sky, of the arc spectrum and of the photometric lamps were obtained so as to be able to have a scale of darkening and hence a scale of intensity for the purpose of calibration. These plates were to be compared later with those obtained during the eclipse. The following day all the plates were developed and we knew then that all our work had been successful.

During the period of totality several other observations of a general nature had been carried out. We had trained not only our personnel but also some of the local people, among whom were schoolboys of the school in Sara, to make drawings of the shape and extension of the corona. Observations were made of the shadow bands which moved rather rapidly from the north east. As is well known the shadow bands, which can only be seen immediately before and after totality, consist of bands of light and shade rapidly alternating, which appear to move very rapidly on the ground in wave-like motions preceding and following the total shadow. This is a phenomenon which is formed in our own atmosphere, probably produced by partial or total reflections and effects of refraction between layers of air of various temperatures which ascend and descend continually in our atmosphere.

In the days which followed we were very busy in dismantling and packing all our instruments in preparation for our return journey. We were also invited to a festival at Novo Povrovka, another small centre about six miles from Sara, to mark the end of the seasonal harvest, at which many of the peasants were present. On the morning of June 25th our coach was once again coupled to the daily train to Orenburg and from there to Kuibiscev (Samara). There we said good-bye to the coach which had been our home, and we continued our journey by river on the Volga as far as Gorkij (Nijni Novgorod) and from there by train to Moscow. From Moscow we went to Leningrad to visit the Pulkovo Observatory. There we had the opportunity of seeing some of the results obtained by our Russian colleagues during the eclipse. From Leningrad the expedition came back to Italy returning on July 12th.

Several months of accurate measurements with special instruments, such as microphotometers and others, were required before the scientific value of the observations taken at Sara could be assessed. The corona (Plate 11) had been extremely bright and in consequence the general darkness of the sky during the two minutes' duration of the totality was not very great, about the same as that experienced at night at the time of full moon. Some very beautiful coronal plumes were visible to the naked eye, in particular an extremely bright one in the direction of Mars and Venus which were less than three degrees from the Sun. The shape of the corona as seen visually and as it appeared on the photographs, was fairly regular in its distribution around the disc of the Sun, in spite of some of the longer plumes. This shape of the corona had been foreseen because the Sun, at the time, was approaching the maximum of its 11-years' cycle. We may here recall the fact that when the Sun reaches its minimum of activity, the corona appears very low at the poles of rotation of the Sun and very elongated at the equator, with very long and regular plumes. The shape of the corona appears therefore to be linked to the regions of solar activities which, in the form of sunspots and prominences, are limited to an equatorial region at the time of minimum activity. At the time of maximum activity, on the other hand, these solar phenomena are distributed over a much wider region almost reaching the poles.

We can say, and this particular eclipse confirmed this very clearly, that single prominences have a very important influence on the shape of the corona and its plumes. In fact at the base of the latter there appears, almost always, a prominence projecting from the Sun's disc. At the poles the coronal rays are very low and curved, as if they were influenced by a magnetic field, similar to that existing on the Earth. Thus we can imagine the globe of the Sun enveloped by coronal rays which are relatively short and curved as if they were repelled by the poles of rotation. At a distance of about 30 degrees from the poles, there are often found long plumes corresponding to prominences, and almost as if these plumes were the continuation of the prominences. The length of the plumes in such cases seems to be linked with the intensity and dimensions of the prominences themselves. The corona, which is a very large envelope surrounding the Sun, consists of very tenuous gases. The spectroscope used during eclipses has revealed that these gases are almost the same as those in the terrestrial atmosphere, only that in the case of the Sun the gases are much more rarefied and in such conditions of excitation as to be able to emit that very special light of a silvery green colour which makes the corona appear so beautiful and impressive. Measurements of the plates made by means of a microphotometer indicate that the intensity of the corona varies with the distance from the Sun and also show how the corona is distributed around the Sun. The difference of intensity between the images obtained in violet and yellow-red light, gives us approximately the temperature of the corona.

The bright emission lines visible in the spectrum of the corona tell us not only the nature of the elements which form the corona, but also their distribution in it. The very well known green line, due to iron atoms in a high degree of ionization, was extremely intense during this particular eclipse. Finally the flash spectrum with its emission lines, supplies us with similar data concerning the chromosphere. Extremely interesting comparisons can be made between these results and those obtained during previous eclipses. The general examination of the plates suggests that the chromosphere is like a rough ocean or even a stormy ocean. There is, however, an important difference. While the ocean on its surface has the same temperature in points relatively close to each other, in the case of the chromosphere the difference of excitation, and hence of temperature and radiation, must be extremely high. In the chromosphere, strong and violent vertical and horizontal currents must dominate.

This particular eclipse held a record for the number of expeditions distributed along the whole of the path of totality and for the good conditions of the sky. In fact only at Kustanai, at about 300 miles from Sara, did the bad weather prevent a French team from making any observations. Similarly, failure on account of bad weather occurred also on the island of Hokkaido, where an English team had set up its station. All the other expeditions were able to carry out their programmes as planned and with good results. Tikhov obtained some beautiful photographs of the corona with four cameras, that is to say photographs in four radiations of various colours. Novakova, with the Czech expedition, took a very good photograph of the flash spectrum using a special instrument with a moving plate, particularly suitable for the determination of the height above the level of the photosphere, of the numerous gases which compose the chromosphere. Perepelkin of the Pulkovo Observatory at Ak Bulak, with an objective prism transparent to the ultraviolet, succeeded in obtaining beautiful photographs of the chromospheric arcs immediately preceding and following the third and second contacts. These plates showed clearly the constitution of prominences and how some of these radiate in the ultraviolet much more intensely than others, thus indicating that they must have a temperature much higher than usual.

The American expedition, which was also located at Ak Bulak, carried out a considerable number of spectroscopic observations and in addition it also studied the propagation of radio waves during the eclipse. A solar eclipse is a very suitable phenomenon to determine whether the influence that the Sun is known to have on the upper layer of our atmosphere, is due to ultraviolet light or to some other radiation travelling with the same speed as light, or to neutral corpuscles travelling with a much slower velocity. If such velocity is of the order of 600 miles per second, as it is thought to be, the eclipse of the corpuscular beam will be observed on the Earth in places other than those where the optical eclipse takes place and which can be accurately determined. At the same time, knowing the velocity with which the corpuscles travel, it will also be possible to calculate the time lag between the optical eclipse and the eclipse of the corpuscular beam. According to the observations of the American expedition during this eclipse, only the ultraviolet light is responsible for the regular diurnal variation of ionization in the E and F layers of our atmosphere which reflect the electromagnetic waves. These results confirmed those already obtained at the time of the 1932 eclipse.

Judging by the articles and communications printed in the daily newspapers in Russia, the interest of the Russian people, particularly of those living in the path of totality, was very great. We were often approached by people enquiring about the circumstances of the eclipse and even on the research we had planned.

All the investigations carried out by the many expeditions during this and other eclipses, collected together will increase gradually our knowledge about the nature and the life of the Sun.

6 · The Solar Eclipse of February 25th, 1952

The phenomenon known as total solar eclipse has always fired the imagination of mankind throughout the ages. This is one of the most awe-inspiring phenomena which is very rare and which lasts only for a few minutes. The very fact that this phenomenon is rare and that the region of the Earth from which it can be seen is very limited, makes the short duration of total solar eclipses so precious and so eagerly awaited by astronomers. If both the Sun and the Moon, in their motion through space, moved on the same plane and the Sun was further away from us, or the Moon nearer, then we would have total eclipses of the Sun each month, their individual duration would be longer and the regions of the Earth from which the eclipse could be seen would be more extensive. In such circumstances the regularity of the phenomenon would make it as common as sunrise and sunset, or as common as the phases of the Moon.

For astronomers a total eclipse of the Sun is of great interest for several reasons. The ancients, usually the priests, who studied the sky, were able somehow to predict only with approximation the dates of eclipses. They based their predictions on the periodicity of the event. Only later did it become possible to predict the time of an eclipse with such precision as to baffle the laymen. This followed the establishing of astronomy as a science, the discoveries of Kepler and Newton and in particular the foundation of celestial mechanics. In order to reach such precision it is necessary to know the laws which govern the movement of both the Earth and the Moon in space and in particular to know very accurately the position of the Earth in relation to the Sun and that of the Moon in relation to the Earth at every instant both present and future. This would be comparatively easy if in the sky there were only the Sun, the Earth and the Moon. Since, however, there exist other celestial bodies at various distances, these produce perturbations and make the calculations and predictions of astronomers extremely complicated. Nevertheless, with modern means of observation and with the comparative ease with which elaborate calculations can be carried out nowadays, we can reach results which may appear very surprising. Suffice just to say that we can predict the time of contacts of the limbs of the Sun and of the Moon within a second, years ahead, and we can also predict very accurately, within a few yards, the small region of the Earth from which the eclipse can be seen as total.

These results may seem amazing but in reality are not much more difficult to obtain than the achievements of man in the design and construction of many wonderful instruments.

It is therefore very important for the astronomers to check by means of clocks and other accurate instruments that we have available nowadays, the difference between the predicted time and the actual time of occurrence of an eclipse. In the field of classical astronomy, if we can determine accurately the time of contact of the limbs of the Sun and of the Moon as seen from various points in the path of totality of an eclipse, it is possible to obtain directly the difference in longitude between the places of observation and hence reach a better knowledge of the shape of the Earth. The fact that Lyot designed instruments which allow us to observe the inner corona of the Sun and its spectrum at any time, without having to wait for the few minutes of an eclipse, has not reduced the importance of the study of the outer layers of the Sun during an eclipse. This type of investigation can be made in many ways and by means of various instruments. Finally, the recent discoveries which have given rise to the new branch of radioastronomy, open a new field in the investigation of the behaviour, during an eclipse, of radio emissions from the Sun at various frequencies.

The total eclipse of the Sun of February 25th, 1952, was particularly favourable for some of the investigations mentioned above. The path of totality, nearly 66 miles wide, started in the Atlantic, crossed French Equatorial Africa, the Sudan, Arabia, Russia and ended, at sunset, in Siberia. The Sudan promised to be the best place for observations. There the sky is almost always clear and in addition, at the time of the eclipse the Sun would be high in the sky. Moreover, the central line of the path of totality passed only $5\frac{1}{2}$ miles south-east of Khartoum, and therefore it was not necessary to be too far from an inhabited centre or to set up camp in the desert. For this reason Khartoum became very crowded with the expeditions of many countries. Here, about 15 expeditions were to be found, including one from Cambridge, several from the U.S.A., consisting of personnel drawn from various observatories and the American Air Force and Navy, a French–Egyptian expedition, a Swiss expedition, and so on.

The Italian expedition, under the auspices of the Accademia dei Lincei, of the National Council for Research and of the Ministry of Education, consisted of Abetti, Righini and Fracastoro from the Arcetri Observatory, Colacevich from the Naples Observatory and Mrs. Abetti.

There were three ways by which Khartoum could be reached from Italy. The fast way was by air from Rome to Khartoum via Cairo. A second way was by sea to Alexandria and then by train and by boat to Khartoum and the third by sea again, through the Suez Canal and the Red Sea, to Port Sudan. From there a twice weekly train of the Sudanese railways reached Khartoum in 26 hours through the desert and the Nile valley.

The Italian expedition decided to choose this third way, because with three tons of equipment in 27 crates, it was necessary to be certain that the equipment should travel and reach its destination together with the members of the expedition. Although the Italian ships which went to Port Sudan were few and far between, we had the chance of joining a banana boat, the *Algida*, on January 4th. This boat was on a regular service between Genoa and Somaliland, and with the help of the Ministry of the Merchant Navy and of the Lloyd Triestino, owners of the boat, she was re-routed to call at Port Sudan on this occasion. We were soon through the Suez Canal and on January 11th we landed at Port Sudan with all our equipment.

Mr. Trucco, manager of a well-known haulage firm in Port Sudan and Khartoum, and who also acted as Italian consul in Sudan, had made all the necessary arrangements for our journey and for the customs clearance of our equipment. We arrived at Khartoum on January 13th and our equipment followed us within four days. Khartoum lies at the junction of the Blue Nile and the White Nile and, together with the nearby town of Omdurman, forms a large conglomeration of about 400,000 inhabitants. In Khartoum are to be found the office of the Governor, the European houses, the new Legislative Assembly and the cathedral while Omdurman, with the famous tomb of Al Mahdi, is mainly a native centre. We, together with most of the members of other expeditions, lodged at the Grand Hotel in Khartoum, on the left bank of the Blue Nile.

Naturally our first problem was to find a suitable site for our station. The Sudanese Government had already made some arrangements and offered the various expeditions sites in military barracks in the immediate neighbourhood of Khartoum. We were particularly anxious to set up our camp as near as possible to the central line of the path of totality, which as we have already mentioned, was $5\frac{1}{2}$ miles south-east of the centre of the town. When we arrived we found an expedition already setting up camp near the central line of the path of totality. This was an American expedition which consisted of van Biesbroeck from Yerkes Observatory, who had set up a large telescope for the study of the Einstein effect, Evans from Colorado University with his multiple spectrograph and several other members of the American Navy whose programme was mainly connected with the observation of solar radio emissions. Their camp was just outside the town and was called 'Kilo Five', because it was very near to the fifth kilometre of the railway line from Khartoum to Kassala.

With the help of Mr. Trucco and of the Comboniani missionary fathers, we soon found a very suitable site which was less than a mile from the central line of the path of totality. Two miles of desert separated Kilo Five from the village of El Gereif. At the edge of this, and near the left bank of the Nile, there was a large cultivated area which belonged to an American Presbyterian Mission, which years ago ran a school of agriculture here for the natives. At the time we were there, Mr. Ebeid Faig looked after this land. He lived there with his family, in a small house with a garden, well protected by trees which surrounded it. We could not hope for any better place than that oasis in the middle of the desert. We accepted very gratefully the hospitality of Mr. Faig. This position was ideal for the protection afforded by the trees against the sand storms which often occurred there. During all the days which we had left until February 25th, we travelled from the Grand Hotel to El Gereif. We worked at the building of the necessary concrete pillars, the erection of the instruments and of the tents to protect them from the very hot sun. We were helped in our work by Mr. Faig and particularly by Brother Adani, who was in charge of the construction of schools and of the many buildings of the Catholic Mission.

Generally speaking we never saw a cloud in the sky, but the transparency of the atmosphere was affected by a wind which was often strong and carried a good deal of sand from the desert. The very dry air and the very high temperature during the day, which in spite of it being winter reached 110°F. one day, were very trying for the members of the expedition and affected our wooden instruments and plate carriers which had all to be repaired. In this respect being in the oasis was a great advantage, plus the fact that we were very close to the central line of the path of totality.

Many are the problems which can be studied during a total eclipse of the Sun. Our programme consisted basically of a repetition of the observations carried out in Russia in 1936, with necessary modifications and improvements. The 1936 eclipse occurred at the maximum of the 11-years' cycle of solar activity, now we were approaching the minimum, which actually occurred in 1954, and we were hoping to compare the observations taken in different circumstances. The programme of observations which we had planned consisted of: (1) the study of the high and low chromosphere which can be photographed immediately before the second contact and immediately following the third contact, in order to increase our knowledge of those levels which cannot be studied in full sunlight; (2) photographs of the spectrum of the corona in the visible, ultraviolet and infrared regions; (3) photographs of the corona with various filters and polaroids in order to obtain the form and intensity of the corona in various regions of the spectrum and to determine the degree of polarization in its various parts; (4) photographs in infrared light of the outer corona, which is still very little known in this region of the spectrum. For this programme, the instruments we used at the 1936 eclipse could be used again, but new instruments had also to be added. Thus to our 'zoo' (see page 64) we added a 'crocodile', a 'python' and an 'armadillo'.

Let us just recall here the instruments used at the 1936 eclipse. The 'giraffe' was a double spectrograph with equatorial mounting having two prism spectrographs, one suitable for the violet and ultraviolet region of the spectrum of the corona, and the other suitable for the violet and infrared region. The 'elephant' was a wooden spectrograph to which, for this eclipse, a metal head was added carrying on one side the slit of the spectrograph and on the other the plate holder which could be moved rapidly in a vertical direction in order to take several photographs at short intervals of time. The 'gazelle' was a double photographic camera with two objectives, one corrected for the blue region of the spectrum, the other for the visible region of the spectrum and to be used with yellow and red filters and polaroids.

The new instruments that were added consisted of the 'crocodile', a universal spectrograph built by the Officine Galileo and presented to the Ettore Gamondi Foundation at the astrophysical laboratory of Arcetri. This instrument was used with a concave grating of 2·8-inch aperture. Since no other optical parts were used, it enabled us to photograph the spectrum of the chromosphere in the ultraviolet. Equipment similar to that used in the 'elephant' was added to this instrument so that the photographic plate could be displaced vertically very rapidly. The 'python' consisted of an optical system with central symmetry designed by Colacevich and which had a focal length of 1·6 inches ($f/0·5$) and a flint prism of 60°. The 'armadillo' was a double photographic camera with short focus and two objectives of 3·2-inch aperture and 12-inch focal length. This instrument was mounted on the middle of the declination axis of the 'giraffe'.

Only the 'giraffe' and the 'armadillo' were equatorially mounted so that all the other instruments required coelostats. These are mirrors which are driven by clockwork and reflect the light of the Sun onto the objectives of the instruments which were mounted horizontally on specially built piers. The 'zoo' was completed by various other ancillary equipment such as batteries, chronometers, photographic cameras and so on.

We needed all the 40 days which elapsed between the arrival of our crates at El Gereif and the actual day of the eclipse. We had to build the piers, test and align the instruments. To make a dark room for our photographic work we had to adapt and modify a little mud hut which existed in the yard of Mr. Faig's house. By means of astronomical observations we determined approximately the co-ordinates of our station. These co-ordinates were found to be 32° 34·8′ east of Greenwich for the longitude and 15° 34·2′ north for the latitude.

The nearness of our station to the central line of the path of totality, made it possible to point both the slits of the 'crocodile' and of the 'elephant' tangentially to the limb of the Sun for the second and third contacts in order to obtain the flash spectrum.

We were not unduly worried about the chance that the sky might be cloudy on the day of the eclipse, but what did worry us was the ever present threat of wind which might bring with it clouds of sand from the desert and in this way obscure the whole sky. On February 22nd we had such an occurrence. A very strong north-east wind raised so much sand that the Sun became hardly visible. Fortunately the weather conditions improved next day and February 25th started fairly clear, although a few faint clouds and a light wind were present. In the days preceding the eclipse we rehearsed all our duties very often. The members of the expedition were responsible for the various instruments and for the observations which were to be made during the eclipse. Additional help was enlisted among local people. The duties were distributed as follows: Abetti with Father Mosna were to work with the 'giraffe' and the 'armadillo', Colacevich and Father Totoni with the 'gazelle', Righini and Father Orlando with the 'crocodile', Fracastoro, Brother Adani and a Sudanese camp guard with the 'elephant'. Mr. Faig was to be responsible for the 'python'. To Mrs. Abetti was left the task of calling out the seconds marked by a metronome, starting from 60 seconds before the beginning of the totality and counting up to 200 seconds, since the duration of the totality was to be 189 seconds.

At the time of the first contact the conditions of the sky were fairly good, although there was a considerable amount of diffusion of light, possibly on account of residuals of sand which had remained in suspension in the atmosphere. At $9^h 44^m 37^s$ local time, the sky began to grow darker until the time of totality (Plate 12). When 10 seconds before the second contact was reached, Righini and Fracastoro started their series of photographs of the flash spectrum.

The author watched the event through the small guiding telescope of the 'giraffe'. The Baily's beads marking the beginning of totality appeared exactly as Mrs. Abetti called the zero second. Our calculations concerning the prediction of the time of totality appeared to be correct within one second, which was the limit of accuracy of time given by our chronometers. At the very beginning of the totality Colacevich began to take the photographs with the 'gazelle' while Father Mosna and Mr. Faig were taking the photographs with the 'python' and the 'armadillo'. Thus during the totality several plates were exposed in order to obtain the spectrum of the corona and direct photographs of the corona with various filters and polaroids. The films and plates of the 'python' and of the 'armadillo' were exposed for the whole duration of the totality. In the last 10 seconds before the end of the totality, Righini and Fracastoro were again busy with the 'crocodile' and the 'elephant' in order to obtain the flash spectrum during those last few seconds.

When an eclipse of the Sun takes place the members of an expedition are the least able, among the spectators, to describe the general aspect of the phenomenon. They have to attend to their instruments during the very precious few seconds available, in addition they are normally under cover and do not get a clear view of the whole sky or of the area surrounding them. Full justice was done to the event by the Press which printed long articles on the subject. We can however state that the inner corona appeared to be very bright, while the outer corona, whose plumes were not very large, did not extend very far. A very conspicuous red prominence was visible at the west limb of the Sun for the whole duration of the eclipse, because the disc of the Moon was not large enough to occult the prominence. To the left of the Sun, Mercury was visible and a little further Jupiter, while on the right Venus was shining very brightly. We cannot say that during the three minutes that the totality lasted, the sky was as dark as at night. On the contrary, remembering other solar eclipses, we had the impression that on this occasion the sky was a little lighter, very much like a sky illuminated by a full moon. This probably was due to a considerable amount of diffusion of light in the atmosphere.

Immediately after the end of the eclipse we carried out the work of photometric calibration by exposing to a source of light plates of the same batch as those used for the eclipse, in order to be able to determine from them the characteristic curve for each plate. From these curves could be obtained the intensity of the light of the corona or of its continuous spectrum, as well as of the bright and dark lines. Once the development of all the plates was completed, part at the camp and part in a more convenient dark room at the Catholic Mission of the Comboniani Fathers, we began the work of dismantling and packing all our equipment. By March 5th everything was ready and the cases were forwarded to Port Sudan.

The results which we obtained from all the plates exposed during the three minutes of the totality, were published in due course in various papers presented to the Accademia dei Lincei.

The corona appears to change its shape periodically according to the 11-years' cycle of the Sun. Lockyer has called the shape of the corona 'polar' type at the maximum of solar activity, 'intermediate' between maximum and minimum and 'equatorial' at minimum. The corona observed during this particular eclipse (Plate 12) appeared to be of the intermediate type. This was as we expected since the eclipse occurred at a time between the maximum and the minimum of solar activity. The two most extended plumes were in the northern hemisphere at about 25° of latitude on the east side of the limb and at 30° of latitude on the west limb. In the southern hemisphere another plume was visible at 45° of latitude on the western side, while on the eastern side the corona appeared rather low.

The regions where the plumes appeared are those which are usually occupied by prominences in the years between the maximum and minimum of activity. It was rather surprising, however, that this time there was no correlation between prominences and plumes as there had been during the eclipse of 1936. Indeed, around the very bright prominence which could be easily seen with the naked eye and which occupied a position angle of 269°, from the north pole of the Sun, the corona appeared almost as if it were absorbed by the prominence itself and was very faint in the immediate neighbourhood of that position.

The spectra in the ultraviolet, in the visible and in the infrared regions presented many bright lines of the corona and of a prominence at the eastern limb of the Sun, near the point of the third contact. It is to be noted that the green emission line of the corona, due to a highly ionized iron atom (Fe XIV), was situated where the corona was observed in the photographs to be rather faint. The flash spectrum showed a great number of reversal lines which helped us in defining the physical conditions of the observed layer of the chromosphere. The small dispersion spectrum obtained with the 'python', indicates that the emission lines visible in it, and which extend to about $1\frac{1}{2}°$ beyond the limb

81

of the Sun, are not lines reflected because of diffusion in our atmosphere but rather are lines which originate in the outer corona. It seems, therefore, that there exists a faint but almost continuous emission from the coronal gases, which develops from the base of the Sun and reaches up to the extreme limit of the corona itself. It is possible that the whole corona is enveloped in a tenuous gas cloud which emits radiation also in its more external part and consists mostly of hydrogen. This is something that until then had not been known. Perhaps this particular result may be confirmed at the time of some future solar eclipse, and may open new fields of research.

In the days which elapsed before the eclipse, several meetings were held at the Gordon College by the members of the many expeditions. At these meetings the various scientists discussed their programmes and the instruments they were going to use. It is safe to say that thanks to the very favourable meteorological conditions, every expedition carried out completely the tasks they had set themselves. From the very large number of observations, considerable knowledge was acquired in the field of solar physics. Van Biesbroeck was kind enough to show us his plates measuring approximately 16×19 inches, which he had secured in his effort to verify the Einstein effect. In them we could see the Sun eclipsed, surrounded by the corona and plumes, and near the corona and in the whole field of the plate there appeared countless stars down to the ninth magnitude. The same region of the sky was photographed again, this time at night in the following July. The comparison of these two sets of photographs enabled the astronomers to measure the amount of the Einstein effect produced by the gravitational field of the Sun.

Great interest in all that was happening around them was shown by the more educated among both the natives and the Europeans. At our camp we had our full share of visitors, among whom were Monsignor Bini, Bishop of Khartoum, and both the Comboniani Fathers and the Nigrizia Sisters with large groups of their respective pupils. We held several lectures in their schools and in the garden of the Catholic cathedral, on the significance of the phenomenon which was to take place. Among the less-educated people the rumour was rife that the astronomers who had gone to Khartoum, were conniving with the devil to arrange that the eclipse should take place in the Sudan rather than in their own countries. In addition it was popularly believed that it was dangerous to look at the eclipsed Sun. As a result all women and children were locked in their own houses by the men, in order to avoid this great danger.

7 · The Mysterious Solar Corona

'The Moon sometimes obscures the whole Sun, but only for a very short time, nor is it large enough to prevent the appearance of some brightness around the rim of the Sun, with the result that the darkness is not completely black nor even completely dark.' With these words Plutarch describes that halo of light visible around the Sun at the time of a total solar eclipse. The ancient observers have always called attention to this halo of light to which has been given the name of corona and which was thought by them to be due to a lunar atmosphere. Modern investigations have shown beyond all doubt that the corona is a phenomenon which belongs to the Sun and is a real gaseous envelope which is extremely tenuous and surrounds it.

The early photographs of a solar eclipse in 1860 by Father Secchi and De la Rue were instrumental in establishing that here was a phenomenon closely linked to the Sun. Since then photography has made great progress. Indeed it can be said that nowadays photography in astronomy has almost completely, in many cases, taken the place of visual observations. If we remember that the longest possible duration of a solar eclipse is seven minutes, but more often only two or three minutes, it is easy to understand that not even an expert artist has the time to reproduce all the peculiarities of the corona. How could we make then a precise study of the shape of the corona on a record which could be very subjective? This is where photography comes to the rescue.

The study of the corona is very important; it is not surprising therefore that efforts have been made to design equipment which would enable astronomers to observe the corona at any time without having to wait for the very short times available during a total eclipse. Secchi and Tacchini attempted to solve this question. They realized that it was most important to have very excellent conditions of seeing, and they thought that perhaps in Sicily, on Mt. Etna, the conditions might be suitable since, as Father Secchi said, 'the sky is of a wonderful purity. That of Rome, compared with it appears hazy and dusty.' The idea was taken up, many years later, by Hale and Riccò and unsuccessful attempts were made, with special instruments, from Mt. Etna.

The great difficulty lies in the fact that the corona is not very bright when compared with the Sun. Parts of the inner corona are of the same brightness as that of a full moon, which is 600,000 times less bright than that of the surface of the Sun. Near the edge of the Sun the light from the photosphere is so intense that it is extremely difficult to photograph the corona in the full light of the Sun. Nevertheless, improved modern techniques enabled Lyot to design and build an ingenious and simple instrument which can be used at any time to obtain photographs of the corona in full sunlight, or at least of those parts known as the inner corona, because they are very near to the chromosphere. The results obtained were remarkable.

During the summer of 1930, Lyot was able to obtain, by means of his 'coronograph', the first direct photograph of prominences and of the inner corona. His coronograph was set up at the Pic du Midi in the Pyrenees, at nearly 9,500 feet above sea level, where the purity of the atmosphere was exceptional. In the years which followed, the coronograph was improved and in addition Lyot designed monochromatic filters. All these new developments made it possible to study continuously the inner corona and its monochromatic radiations.

Total eclipses of the Sun, however, have not lost their value for astronomers. It is during the eclipses that we can study the more external layers of the Sun and the complete corona, at great distances from the Sun. The region of the Earth from

where the totality is visible is rather small and therefore astronomers have to travel to distant shores, sometimes to very un-
comfortable places. If the path of totality covers the oceans then
precise observations from ships are out of the question. Perhaps
the American astronomer S. A. Mitchell is the one who has
observed the greatest number of solar eclipses. He has observed
nine eclipses of the Sun, and in order to do that he has travelled
90,000 miles. According to him, during these nine eclipses, he
observed the eclipsed Sun only for a total of 18 minutes.

Even those who have never actually seen a total eclipse of the
Sun must have seen drawings or photographs, read descriptions
of the phenomenon and even followed by television the eclipse of
1961. They know how the phenomenon occurs. The Sun slowly
disappears completely behind the disc of the Moon, which can
then be seen surrounded by the solar corona which consists of
plumes of various length and curvature, reaching far away to
great distances from the Sun. At the base of most plumes are to
be seen, here and there, bright prominences of a vivid red
colour.

When we study the general shape of the corona, as it has been
observed during the eclipses of the last 50 years or so, we find
that it varies from one eclipse to another. It was possible to
establish a close relationship between the shape of the corona
and the phase of the 11-years' cycle of the solar activity. When
the activity is at a maximum, and therefore we see on the surface
of the Sun large numbers of sunspots and eruptions, the corona
is almost circular and has the appearance, as Janssen remarked,
of an enormous dahlia, or of a halo with plumes which are more
developed at the higher latitudes. At the time of minimum, on
the other hand, the corona is generally less extensive, although
at the latitude of the equator long plumes are present which
gradually merge in the diffuse light of the sky. Langley, observing the eclipse of the Sun in 1878 from a height of about 13,000
feet above sea level, was able to trace these plumes to a distance
of nearly 12 times the diameter of the Sun.

We know that, like the Earth, the Sun rotates around its own
axis. When we observe the corona at the time of a minimum of

the solar cycle, we can usually determine, without difficulty, the position of the axis of rotation, because in the neighbourhood of the poles of the Sun we detect plumes which are relatively short and curved, very much like the lines of force of a magnetic sphere near its poles. When, on the other hand, we observe the corona at the time of a maximum of the solar cycle, the plumes are so numerous and often overlap each other so much that they may mask completely, or at least distort, the polar plumes. These observations have led astronomers to classify the various types of coronae. At the time of maximum activity of the Sun, it is known as a polar or irregular corona; between maximum and minimum the corona is of intermediate type while at the time of minimum activity of the Sun, the corona is of equatorial type.

It is natural therefore to suspect that the shape of the corona is directly dependent on the internal conditions of the Sun, as well as on its periodic phenomena which appear not only on its surface but also on the higher parts of its atmosphere. The great disturbances which occur in the lower part of the atmosphere of the Earth, on account of the varying amount of heating from the Sun, do not generally extend to great heights. This, however, does not exclude the existence of disturbances, periodic and non-periodic, in the stratosphere. Part of these disturbances are no doubt produced by the Sun but part may well be due to phenomena which take place nearer to the surface of the Earth itself. On the Sun, sunspots and the eruptions which we observe in various forms, either projected on the disc of the Sun or rising from the solar limb to great heights, are periodic phenomena. If we remember that sunspots appear at relatively high latitudes at the beginning of a solar cycle and then gradually descend towards the equator, then we have a possible explanation for the variation of the shape of the corona. Nevertheless, this explanation is not completely satisfactory since the coronal plumes in the neighbourhood of the poles do not appear to have any link whatsoever with the sunspots. On the other hand, if we take into consideration the prominences and their periodic migration in latitude during the solar cycle, then we have better results.

87

Unlike sunspots, prominences are not limited to a belt between the equator and latitudes of 30° north and south, but may reach the poles. At the time of minimum solar activity, prominences are found with greater frequency at a mean latitude around 50° north and south. About two years later, the prominences divide themselves into two distinct zones. One descends gradually towards the equator while the other ascends slowly towards the poles of the Sun.

If the plumes are to follow the prominences, then they ought to migrate from the medium latitudes both towards the equator and towards the poles. This is actually what happens. About two years before the maximum of the solar cycle the high-latitude prominences reach their maximum activity and at the same time the highest latitude and the corona tends to acquire a polar type form. This form varies slightly according to whether the solar activity increases from the minimum to the maximum, when the prominences are already divided into two zones of high and low latitude, or whether it decreases from maximum to minimum and the prominences are found only at medium or low latitudes. Interpreting the phenomenon in this way helps to explain the fact observed that the maximum frequency of prominences at high latitude, which is almost coincident with a corona of polar type, is out of phase with the maximum of the sunspots. Something similar occurs at the time of minimum.

We must always remember that the phenomenon described does not happen in a precise and regular manner, as for instance the alternation of day and night or the succession of the seasons on the Earth. The only regular phenomenon which occurs in the case of the Sun, is its rotation around its own axis, which takes place in approximately $25\frac{1}{4}$ days except for the variations according to the latitudes considered. Although the manifestations of the solar activity follow, on average, a periodic cycle of 11 years, they do not repeat themselves with the precision of a mathematical law. Indeed the 11-years' period itself may be subject to variations of a year or two. If we wish to make a comparison with events on the Earth, let us take as an example the forecast of meteorological phenomena. Although these follow a general rule, nevertheless they vary very widely as is known only too well, and for the time being at least, cannot be predicted over a long period of time. The same happens on the Sun and hence in the case of the corona we may well meet with exceptions and variations from the general rule described above, in relation to the presence or otherwise of prominences at the solar limb.

In order to be able to unveil some of the mysteries presented by this extended envelope of the Sun, it is necessary to know of what elements it consists. Spectral analysis will help us in this task. Some years ago, in the inner corona, emission lines of various intensities had already been discovered. To date 26 lines in all have been detected, ranging from the ultraviolet to the infrared. The identification of these lines was not an easy matter for the astrophysicists. In fact in a first attempt some lines were attributed to an unknown element which was called 'coronium'. Gradually, however, following the progress of spectroscopic work in laboratories, on gases of the stratosphere which gave rise to aurorae and finally on the spectra of novae, it became possible to solve this particular question. In 1945 Edlen was able to prove both theoretically and experimentally, that highly ionized atoms of iron, nickel and calcium were responsible for the majority of the emission lines of the corona. By highly ionized atoms we mean atoms which, because of the high excitation to which they are subjected, have lost a few of their more external electrons. The high ionization of the atoms points to the fact that a very high temperature estimated to be of the order of a million degrees, must exist in the corona. Actually it is a question of a temperature of an extremely rarefied gas, having the atoms in a state of great excitation and turbulence, which is not to be confused with the temperature of radiation of the photosphere and chromosphere which is only of the order of 10,000 degrees. One thing is certain, and that is that the processes which take place and produce the observed phenomena and temperature have not yet been explained satisfactorily. One theory is that as the chromosphere is formed and heated by the photosphere, similarly the corona is produced by electromagnetic phenomena of the chromosphere.

8 · The History of the Sun from 1933 to 1965

Certain classes of natural phenomena follow well-defined mathematical laws which enable man to predict these phenomena in the future with great accuracy. This is true unless unexpected cataclysmic conditions occur which cannot be foreseen and which are very unlikely over short periods of time, such as those considered by man. Other phenomena, although they follow a given behaviour, are very irregular and they do not seem to follow precise laws, or at least we must admit that so far we have not been able to discover laws which would satisfy the occurrence of these phenomena. Typical examples are the meteorological phenomena, which are difficult to predict and when predicted can only be expressed with a certain degree of probability over short intervals of time of one or two days. The Sun, on the other hand, presents several phenomena, but so far, we have not discovered the laws which rule them. Nevertheless, the fact that these phenomena are repeated with a certain regularity makes it possible for us to predict their occurrence with a fair approximation without running a serious danger of falling into great error.

The 11-years' cycle of the Sun is well known, and as described earlier (see page 52) consists of the repetition every 11 years of a series of storms of various intensities, which disturb the surface of the Sun. These storms appear in the form of sunspots and of eruptions of incandescent gases, mostly hydrogen, calcium and helium, which are released from the interior of the Sun and are seen either as bright patches on the disc or, when they occur near the limb of the Sun appear as huge flames which we call prominences. Solar storms, like terrestrial storms, do not appear regularly either in the same place or at given times, nor do they appear punctually every 11 years. One thing is certain and that is that the Sun, from a period of relative quietness, in about five years reaches a maximum of activity, only to return to its original quietness in the following six years. Throughout the many years when these observations have been made, starting from the times of Galileo, it appeared that the period of the solar cycle is not always of 11 years. Periods one or two years longer or shorter than the average 11 years have occasionally been detected.

Storms which take place on the Sun can be of different intensities and can also affect regions of various extent and latitudes, according to the stage of the cycle at which they occur. If we could possibly compare terrestrial phenomena which take place on a solid body, with the solar phenomena, occurring on a gaseous body, we might say that the sunspot represents a crater, while the eruptions are violent emissions of various incandescent gases, which are often ejected from the crater itself, or more often, from its neighbourhood.

Since the days of Galileo to our time, nearly 30 solar cycles have elapsed of an average of 11 years each and not all of them have shown the same intensity.

Only four years after Hale discovered, in 1908, the existence of magnetic fields related to sunspots did a new important discovery follow. These magnetic fields have always opposite polarity in the two hemispheres of the Sun and, what is more, they change their polarity from one 11-years' cycle to the next. Because of this, the duration of a complete solar cycle must be considered to be twice the length, namely 22 or 23 years instead of approximately $11\frac{1}{2}$ years. In fact, after 22 or 23 years the Sun returns to the beginning of a new cycle, having exactly the same conditions as at the beginning of the previous cycle. Hale has shown the existence of a general magnetic field of the Sun similar to that existing on the Earth, in addition to the intense magnetic fields which develop around the sunspots. Later, further observations appeared to cast some doubts upon the results obtained by Hale and his associates. In recent years, however, Babcock, father and son, with a very ingenious instrument used in conjunction with the solar towers of Mt. Wilson, confirmed that a general bipolar magnetic field exists on the Sun. The intensity of this magnetic field was less than that determined by Hale in 1912. At the time of the observations by Babcock in 1933, the magnetic field of the Sun appeared to be of 2 gauss, and with a polarity which was opposite to that existing on the Earth. Furthermore, the magnetic field of the Sun appeared to be subject to irregular fluctuations both in intensity and size. These polar magnetic fields of the Sun originate in those regions from which emerge the typical curved polar rays of the corona which are such a conspicuous feature of photographs taken during a total eclipse of the Sun.

If we want to describe the recent cycle of 22 years (1933–1955) we can say that in October 1933 we had a minimum in the number of sunspots. This quantity is known as the 'relative number' or the 'Wolf number', and is a conventional quantity which takes into account the number of groups as well as the number of individual spots which appear on the surface of the Sun (fig. 6). Now, in October 1933, the Wolf number was, on average, six for each day and all the spots appeared in the neighbourhood of the equator, both in the northern as well as in the southern hemisphere. After that year sunspots began to appear at higher latitudes between 30° and 40° both north and south of the equator. The number of sunspots gradually increased until it reached a maximum of 114 in 1937. During this cycle solar disturbances of great intensity affected, as usual, terrestrial phenomena, in particular causing large disturbances of the terrestrial magnetic field and producing aurorae which were observed at latitudes as low as Italy two or three years after the maximum activity of the Sun, when the frequency of solar disturbances was in fact beginning to decrease. In May 1944 another minimum was reached with a Wolf number of 10, with a few spots in the neighbourhood of the equator. The duration of this particular cycle then was of 10½ years, at least as far as the phenomenon of sunspots is concerned.

After 1944, the number of sunspots began to increase again and fairly rapidly this time. In 1946 and in 1947 there occurred some very violent manifestations of solar activity such as had

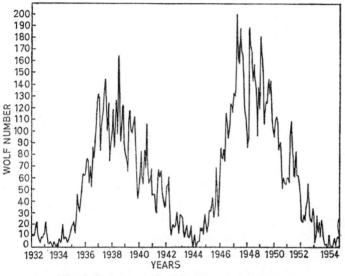

Fig. 6. Sunspot cycles for the period 1932–1954

been rarely observed during the last three centuries of regular telescopic observations. The most intense manifestation took place on July 25th, 1946, when a very large group of sunspots, the largest ever observed, was almost completely covered by very intense flares, consisting mostly of hydrogen and helium gases.

When we mention flares or chromospheric eruptions, our mind automatically thinks of terrestrial eruptions in which violent chemical combinations of molecules occur. On the Sun, on the other hand, because the temperature is so much higher, we are in the presence of atoms in a very high state of excitation which emit light because of the transition from one level to another of the electrons surrounding the nucleus. Therefore, in the case of the Sun, when we use the term 'eruption' we intend to give only an indication of the violence and speed of the

93

phenomenon, which in effect is on a much greater scale and which is very different from the phenomena which occur on the Earth. Simultaneously with this particular violent solar pheno-menon of July 1946 disturbances in radio communications were recorded on the Earth and about 25 hours later, strong distur-bances of the terrestrial magnetic field were also recorded. In addition, intense aurorae were observed both in northern Europe and North America.

During the year 1947 occurred the maximum of the 11-years' cycle 1944·2–1954·3, which, as can be seen, in this case was only of 10·1 years. This was the fourth consecutive cycle which had a shorter duration than the average 11·5 years. The Wolf number for the maximum of solar activity in 1947 reached the very high value of 152. Since the year 1749, when regular and reliable ob-servations of sunspots began, only the maximum of 1778 reached a higher value of 154.

Because in the early days the only observations of solar phenomena were limited to the observations of sunspots, no valid comparison can be made between the two values mentioned above. Nevertheless, what is certain is that the maxima for 1778 and 1947 were exceptional when compared with the other years. After 1947 the activity of the Sun decreased, at first slowly until in 1953 and 1954 there were many days, and even months, dur-ing which the Sun did not show either a spot or any activity worth recording. If we want to find a similar period of quietness in the activity of the Sun we have to go back to 1810 and 1811.

In the sequence of the various 11-years' cycles, it is a regular occurrence that before the end of one cycle there appears some activity belonging to the new cycle. Thus already in August 1953 a very small spot was observed at high latitude and this was the forerunner of the new cycle 1955–1966. After that other small groups of spots at a latitude of about 28° in both the northern and southern hemispheres began to appear. They were short-lived as if they had no energy to develop fully and to last longer. The magnetic polarity of these spots confirmed the general rule. The polarity was of an opposite sign to that of spots of the old cycle, some of which were still in existence at a latitude of ± 5°.

More exactly single spots, or the leading spots in the bipolar groups in the northern hemisphere, had a north or positive polarity, namely the same polarity as the north terrestrial pole. The opposite happened for spots in the southern hemisphere.

There were also two other manifestations which proved that the new cycle was beginning, that is the presence of faculae and filaments in the neighbourhood of the Sun's poles and an awakening of activity in some regions of the solar corona. The corona, photographed in Sweden during the total eclipse in the year 1954, was of the equatorial type. This type of corona, which is reminiscent of the appearance of the lines of force surrounding a magnetized sphere, is characteristic of the period of minimum of solar activity.

The activity of some of the regions of the corona, around the medium latitudes where sunspots are evident, can be followed by means of coronographs situated at high-altitude stations, as for example that of the Pic du Midi in the Pyrenees. At this observatory, towards the end of 1954, were observed important coronal plumes at latitudes which were generally higher than those occupied by sunspots.

The 1944–1954 cycle, which lasted 10 years, was followed in 1955 by a new cycle which showed immediately an exceptional intensity. It was as if the Sun had been aware of all the preparations that were made on Earth for the International Geophysical Year and wished to co-operate with some exceptional manifestation of activity!

One of the main reasons for the selection of this particular time for such a great and important international undertaking was that from the knowledge of the behaviour of solar activity, 1958 was due to be a year of maximum of the 11-years' cycle. As has already been said, the duration of the cycle is not exactly 11 years, nevertheless in 1954 there had been a minimum of solar activity so that in 1958, or thereabouts, a maximum could be expected. Similarly, for the next cycle it could be predicted that the next minimum would occur in 1965. From the experience of over three centuries it was expected that after 1954 the disturbances on the Sun would begin to increase gradually. The cycle develops in such a way that the increase to a maximum is faster than the decrease from maximum to minimum. This is the case with many phenomena of evolution, including, for example, human life.

Without being able, of course, to predict exactly the day or even the month, it was however possible to foresee that four or five years after the minimum of solar activity, a maximum would occur. Therefore as 1958 approached the astrophysicists began very complete and careful observations of all the solar phenomena. These included the presence of sunspots on the solar surface which are often accompanied by large eruptions of hydrogen and other gases and by an exceptional increase of ultraviolet radiation.

The early observations of the Sun were started by Galileo in 1611, but continuous observations upon which statistical analysis can be based were really initiated in 1749. From that year up to the present we have data covering just over 19 cycles, and each cycle has shown intensities of various degrees.

The measure of the intensity of a cycle can easily be done simply by counting the average number of sunspots visible in a day, or, better still, by measuring the area of the Sun covered by sunspots, an area which is always very small compared with the total area of the Sun. This was the method used up to about 70 years ago. In modern times with new techniques and the addition of new instruments, it has been possible to add many other important observations of other solar phenomena which are known as solar disturbances or storms. During the course of the 19 cycles we have records of the diurnal number of sunspots and the year 1778 had the highest number, followed very closely by the year 1947.

When we examine the general trend and intensity of the 19 cycles, we notice at once that there are considerable variations. Some are longer and less intense, while others are shorter but more intense, as if there existed a relationship between duration and intensity. This pattern seems to be repeated with a certain regularity. Mathematicians and physicists tell us that this may be the result of the overlapping of what could be called 'waves' of shorter or longer duration, and which tend to combine together. It is not very difficult, within certain limits, to separate these various 'waves' and to find the primary ones for both short and long durations. Actually this has been done and although we do not, as yet, have definite data it can be said that the Sun is subject to periodic fluctuations. The most obvious of these is that lasting 11 years, and on this are superimposed others, one of 8 years and another of anything between 70 and 100 years.

Without carrying out any complex calculations, we see at a glance that from 1749 there are groups of cycles which are longer and less intense alternating with groups which have a rapid and violent increase and which are of shorter duration and have a greater intensity. Thus the exceptional maximum of 1947 followed that of 1937, which was also of exceptional intensity. In forecasting the next maximum after 1947, it was at once obvious that the solar activity was already beginning to be particularly violent and therefore the next maximum would belong to an intense group.

This actually happened and now we can say that the record reached in the year 1778 has been broken. The diurnal number of sunspots, or the relative number, which is calculated at the Zurich Observatory on the basis of many observations obtained all over the world, was 154 for 1778, 152 for 1947 and 190 for 1957.

The maximum of the nineteenth cycle was reached towards the end of 1958 while in 1956, only two years after the minimum, the activity had already reached a remarkable intensity. During 1957 the Sun continued to have very violent disturbances. In many consecutive days there occurred some extremely rare cases when the number of spots was over 300 and reached even 355 on December 24th–25th, 1957. In 1958 many solar disturbances were still apparent and although the actual maximum was passed, large spots appeared on the disc of the Sun even about January 24th, 1959. That evening, at sunset, an observer at the Cidnea Observatory (Brescia) with an ordinary pair of binoculars was able to see, in the penumbra of one of the largest spots, a very sudden flash of light which lasted about ten seconds. No doubt this phenomenon was a flare which, although normally can only be seen with special equipment, must have been so intense as to be visible almost to the naked eye. This type of event has occasionally occurred in the past.

This exceptional and extraordinary cycle had some remarkable characteristics which could be summarized as follows: (1) The almost total absence of spots of large dimensions as are generally present in normal cycles; (2) the high latitude at which spots were found. In particular one had been observed at a latitude of 50° north while, generally speaking, spots are rarely seen at latitudes higher than 40°.

All these facts may give some indication of the exceptional and turbulent conditions existing in the interior of the Sun. It is from there that the solar cyclones originate, that is to say the spots which, when they reach the surface, are animated by movements of gases which are almost regular. These gases are ejected and they re-enter and rotate around the centre of the cyclones with various velocities. In certain special conditions the equilibrium of the vortex is subject to small shifting movements on the surface of the Sun and can be disturbed in such a way as to break up at many points in a chaotic manner producing various phenomena such as flares, filaments and prominences accompanied by their magnetic fields.

After the 1964–1965 minimum the twentieth cycle begins. All we can say about this cycle is that it will cover the period 1965–1976, but at present we are not in a position to make any predictions concerning its intensity. If we rely on the evidence of past cycles we must now expect a series of cycles of a lower intensity.

9 · The Radio Sun

The human eye and what could be called the photographic eye, assisted by powerful optical aids, have succeeded in discovering and observing many solar phenomena. These phenomena reveal to us, up to a point, the physical nature of the Sun and the development of its life which is at times very stormy and at other times quieter.

The discovery that celestial objects emit radio waves (see page 30) of various wavelengths has opened a new and extensive field of investigation for astronomy and has increased man's capability of investigating the mysteries of the universe.

Our terrestrial atmosphere is a serious obstacle in the study of the radiations emitted by the Sun because it absorbs a considerable part of this radiation. It is true that nowadays we can launch rockets equipped with suitable instruments well above the terrestrial atmosphere and observe these radiations which normally do not reach the observer on Earth. These experiments, however, are only beginning, and the equipment which can be carried by a rocket is of necessity very modest, therefore observations from the Earth have still to be carried out.

Radiations of very short wavelength, such as the X rays as well as a great part of ultraviolet radiation, ionize the gases of the upper atmosphere so that their energy is absorbed at those heights and the radiation cannot reach us on the Earth. At a wavelength of about 0·3 μ the atmosphere begins to be transparent both for photography and for the eye. Radiations from this wavelength to that at the beginning of the infrared can be observed. Beyond the beginning of the infrared, absorption starts again, this time produced by the presence of molecular oxygen in water vapour and carbon dioxide. When we reach wavelengths of the order of a few millimetres, the atmosphere is once again transparent, and this transparency extends up to radiations of a wavelength of 10 metres. Longer wavelengths are stopped by the rarefied atmospheric layers which are found at heights between 60 and 250 miles. As the radio waves of a long wavelength emitted by terrestrial transmitters are reflected to Earth by the ionosphere, so the cosmic radio waves are reflected back into space by the same layers (fig. 5).

It is almost unbelievable how in a comparatively short time powerful radio telescopes were developed by the efforts of radio engineers more than by that of astronomers. These radio telescopes make use of aerials of various types which in some cases consist of very complicated arrays and are used to receive radio waves emitted by the Sun and the stars. At present there are many observatories scattered over the face of the Earth which have radio telescopes capable of following the Sun in its apparent diurnal motion and of scanning the sky in their quest for the celestial regions which emit radio waves.

Since the Sun is a star so relatively near to us, it is natural that the observations by means of radio telescopes should be of particular interest to us. Already valuable results have been obtained. As a result of these observations we can now classify radio waves emitted by the Sun into two fundamental types. The first refers to radio waves emitted by the quiet Sun, that is to say when its surface does not show any particular phenomenon. The second refers to radio waves emitted when the Sun is disturbed, namely when its surface shows phenomena such as flares, sunspots and prominences, which can be detected by visual and photographic methods. When the Sun is quiet, our receivers and recorders, connected to the aerials, reveal radio waves of various wavelengths which appear to be emitted by the whole hemisphere visible to us, and which seem to be constant in time, at least as far as we know at present.

Following the practice adopted in radio communications, radio waves in radioastronomy also are expressed either in wavelengths measured in metres or in frequencies expressed in number of cycles per second. One thousand cycles per second represent a kilocycle per second and a million represents a megacycle per second—abbreviated kc/s and Mc/s respectively.

When we measure the energy of the solar photosphere emitted in the optical range of the spectrum, we obtain for the surface temperature of the Sun 6,000°K. On the other hand, we obtain different results if we measure the energy emitted in the radio spectrum. In fact at wavelengths shorter than 3 centimetres (10,000 Mc/s) the temperature is much higher, of the order of 20,000°K and at wavelengths of 1·5 centimetres (200 Mc/s) the temperature is higher still, reaching 1,000,000°K.

From observations of the optical spectrum it has been found that temperatures of the order given above exist in the chromosphere at a height of 6,000 miles above the photosphere and in the corona, above the chromosphere, up to distances of several solar radii from the photosphere. The high-frequency radiation which may escape from the lower layers of the chromosphere and which has therefore both a chromospheric and a coronal component, shows an effective temperature lower than that of low-frequency radiation, which originates only in the corona at a temperature of 1,000,000°K.

When the Sun is disturbed we find emission of radio waves of various wavelengths. Between 3 and 60 centimetres (10,000 to 500 Mc/s) there is a component in the emission which varies slowly within a period of 27 days and which must originate in disturbed regions of the Sun (sunspots and flocculi). Whenever these, on account of the rotation of the Sun, are directed towards the Earth, the emission of radio waves occurs again within a period of 27 days and with the same frequency.

When sunspots, or groups of sunspots, are of great dimensions, that is when their area is greater than a 400 millionth of the solar disc, then we receive from the Sun intense radio waves of wavelengths between 1 and 15 metres (300 to 20 Mc/s). These radio waves can be heard in loudspeakers as 'noise' or can be recorded by special pen recorders.

The radio emissions may last for hours and days and may be accompanied by rapid fluctuations of greater intensity, like 'bursts' of radiation, which may be repeated several times at regular intervals of time. Evidently these are the result of solar storms which take place within and in the neighbourhood of the large sunspots, and also of the energy they develop, as happens, on a smaller scale, in the case of storms in our own atmosphere. The energy of the solar bursts of radiation is calculated to be several thousand times that emitted by the quiet Sun. The intensity variation of the bursts varies with the wavelength and therefore it is a great help to have continuous observations of the intensity in order to define the characteristics of the bursts. By following a group of sunspots during a solar rotation and by recording the arrival of the radio waves, we conclude that the source of the radio emission must be at a much higher level than that of the sunspots, and hence is probably located in the corona.

Another class of radio emission, which is usually called an 'outburst' because of its intensity, is also observed when the Sun is disturbed. These outbursts may last for only a few minutes or even a few seconds. It is almost certain that their origin is to be found in the flares. We have already mentioned (see pages 30 ff.) that flares emit a very intense ultraviolet radiation which has a considerable influence upon the ionosphere and simultaneously we also have a strong emission of radio waves (fig. 7), or outbursts, with wavelengths from 8 millimetres to 15 metres (38,000 Mc/s to 29 Mc/s). Very often an increase of intensity in cosmic rays is observed in association with flares and outbursts. It would appear, therefore, that cosmic rays could well be generated by solar phenomena. Whether these two phenomena are in turn linked with the emission of radio waves, is something of which, so far, we are not certain.

These investigations on the radio emission from the Sun and other celestial objects have only just begun. In the case of the Sun they have confirmed the high temperature which exists in

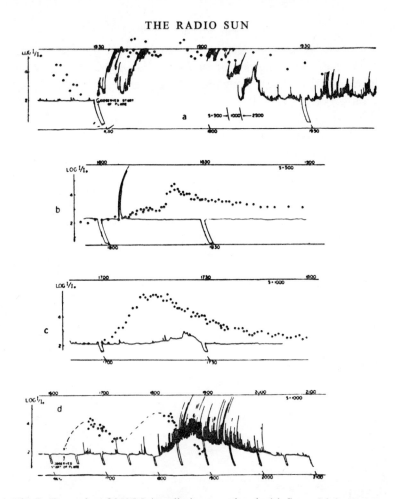

Fig. 7. Examples of 200 Mc/s radiation associated with flares. Light curves for the flares (series of dots) are superimposed on the radio records. (*a*) Major burst May 20th, 1950; (*b*) Minor burst October 17th, 1950; (*c*) Small rise in base level June 22nd, 1949; (*d*) Series of bursts June 1st, 1950. (The horizontal scale represents Universal Time and the vertical scale represents the intensity of flares and of bursts on 200 Mc/s)

the corona and which was already suspected from the study of its spectrum. Moreover, radio observations of the Sun confirmed the existence of a general magnetic field, another question which had already been studied by ordinary optical observations. In addition to all this, the radio observations have supplied us with a knowledge of the electron temperature and density of the regions of the upper chromosphere and of the inner corona which, in the optical range, could only be observed during an eclipse (see pages 86 ff.).

What is of great importance is the discovery of the large amount of energy which is emitted in the radio spectrum by solar storms. By means of radio waves we can now investigate the early stages of the ejection from the Sun of those corpuscles which produce corresponding terrestrial phenomena.

All these developments in this very new branch of solar physics promise new discoveries and a better understanding of that star which sustains life on the Earth.

10 · The Utilization of Solar Energy

If we consider that the energy received on the Earth from the Sun is nearly 100,000 times the electric energy produced in the whole world, it is not surprising that man has tried to capture it and to convert it for ordinary daily use. Already nature itself has done much in this direction if we think of the large amount of coal available on the Earth which is a product of solar energy, or of waterfalls which can be used to produce electricity, since after all rain itself is a product of solar energy.

The direct utilization of the very large amount of solar energy available presents great difficulties because such energy is, as it were, very much diluted in space. For the moment, at least, the use of natural resources such as coal, oil, natural gases (methane) and hydroelectric energy are more economical and more practical. In recent years we have added to these atomic energy which already is taking an important place among the other types of energy and there is little doubt that in the very near future atomic power stations will play an extremely important part in the production of the energy required all over the world. The need for energy is very great indeed and increases daily because of the rapid development of industry. It is therefore obvious that the studies and the investigations which are now taking place in order to find new and more economic means of producing energy are far from superfluous and that in particular, an effort should be made to harness the enormous amount of energy released by the Sun.

The direct use of this type of energy, for the time being at least, has not yet reached a very advanced stage. Nevertheless, particularly in those regions of the Earth where the Sun is more generous and shines almost every day, solar energy can be extremely useful.

The attempts made to produce energy cheaply by using the Sun are not new. According to legend, Archimedes, by using large mirrors, succeeded in setting on fire the fleet of Consul Marcellus off Syracuse. A few years ago, both in France and in Egypt, solar centres were built where water vapour was produced by means of large metal mirrors which concentrated the solar rays on copper cylinders which acted as boilers. The motors which were driven by the steam were used for pumping water. These installations, however, have an efficiency of only about 5 per cent, mostly because of the very low efficiency of the steam engine. In more recent years in the eastern Pyrenees, solar furnaces have been constructed. The great amount of sunshine in that region and the more modern design of mirrors which can follow the apparent diurnal motion of the Sun, succeeded in producing a temperature higher than 3,500°C. at the point where the mirrors focus the light of the Sun. This type of furnace is particularly useful for metallurgic operations, for the firing of ceramics and for obtaining chemical reactions which require a very high temperature. The installations in the Pyrenees, which so far are of a modest size, are very promising and it is hoped that by enlarging them it will be possible to produce an output of 100 horsepower and in this way carry out daily the treatment of hundreds of pounds of various metals.

A more modest use, still based on the principle of concentrating the solar energy by means of mirrors, is that of the solar cookers, and of plants for the distillation of sea water. Various types of these appliances are now available on the market and they can be used, naturally, only in those countries where there is no lack of sunshine.

In the United States, particularly in Florida and in California, where the daily amount of sunshine is high, houses have been built in which in a part of the roof, facing south, there are blackened pipes in which air and water run. These pipes are then covered by panes of glass, just like a greenhouse. The air and water which are heated by the Sun, circulate in the system of pipes just as in an ordinary central heating system and finally are collected in a special cistern. In this manner it is possible to store the heat obtained during the day and to use it for heating the house during the night. Indeed the heat may be collected and stored for many days and therefore also for cloudy and foggy days. From tests carried out it appears that in such houses the fuel necessary to maintain a reasonable temperature in the building and a supply of domestic hot water, can be reduced by more than 50 per cent.

Similar installations are also used for providing air cooling in houses following processes like those adopted in refrigerators. In this manner houses, particularly those in warm climates, can be kept at a constant temperature throughout the year.

Another method for storing solar energy and heating houses has recently been attempted in Massachusetts. In a house situated in a region which is not particularly favoured with clear skies and which was heated for two winters by solar energy, there is a rather large wall about 18 feet high and 36 feet long. This wall receives the light of the Sun and heats a solution of sodium sulphate to a temperature of about 91°F. At this temperature the solution loses its water of crystallization. When the Sun sets and the temperature falls below 91°F., the sodium sulphate which is now dry, in re-absorbing its water of crystallization develops heat which maintains the solution at 91°F. The solution is kept in metal containers which are placed between the walls of the various rooms and the air is made to circulate through the hot containers and the rooms. The sodium sulphate solution is capable of heating the house for about 10 days if there is no Sun, and this period of time has proved sufficient even in a climate like that of Massachusetts.

For many years now production of electricity directly from the heat of the Sun has been obtained by means of thermocouples even if on a small scale. When two small plates of different metals are joined together and one of them is heated while the other is kept cold, a very weak electric current is produced. Photoelectric cells, which nowadays have such wide applications, convert light into electricity. Both thermocouples and photoelectric cells, however, are not really suitable for the production of a considerable amount of electric energy.

In recent years a discovery made by scientists of the Bell Telephone Company in the United States, has led to the construction of a new type of thermocouple which has a very high efficiency. This offers a very useful and economic source of electric energy which can be adapted to many purposes. The element used for these new thermocouples is silicon, which is so widely distributed in nature. The much higher efficiency of batteries made up of these thermocouples makes the use of solar energy a practical proposition. Solar cells of this type have in fact been widely used in artificial satellites.

Part of the solar radiation between the violet and the ultraviolet which penetrates the atmosphere and reaches the Earth has beneficial effects on living organisms. Nowadays it is widely used for therapeutic purposes and indeed even artificial sources have been created. It has been discovered, for instance, that when the skin is exposed to the direct light of the Sun, apart from the tanning effect which is visible, it produces also vitamin D, which is essential for the regular functioning of the human organism. When we do not receive enough solar radiation, the microscopic laboratories of our skin cease to produce vitamin D with the result of a general debility of the organism. Very important results are expected from these investigations.

There is no doubt that those regions which are far removed from the man-made centres of production of energy and which are deprived of natural resources, such as coal, water, etc. but which have abundant sunshine, will, in the future, make wide use of solar energy.

11 · Sun and Earth

Life as we know it on the Earth in all its various forms and manifestations owes its existence to the energy and heat radiated by the Sun. The energy and heat of the Sun is of course not only directed to the Earth but is dissipated in the whole of space and the intensity of it is stronger near this large source and much weaker further away. It is an easy matter to calculate the amount of energy received by Mercury which is the nearest planet to the Sun, and by Pluto which is the furthest and which was discovered only a few years ago. From this information we can deduce that neither on Mercury nor on Pluto, is the existence possible of the type of life to which we are accustomed on Earth and which is impossible at temperatures even a few degrees higher or lower than those prevailing on the Earth. In fact this can be tested even on our own planet. It is sufficient to climb to the top of the highest mountains or approach the poles in order to discover how gradually both animal and vegetable life disappear if it does not cease to exist altogether.

One of the problems which has been amply discussed and studied is that of whether solar energy has undergone any variation in the past or whether it may undergo changes in the future so that conditions on the Earth may change to such an extent that life as we know it may be subject to considerable modifications. Such variations that have been discovered, particularly on the evidence of geological phenomena, are more likely due to the actual position of the Earth in space rather than to changes in the energy radiated by the Sun. Nevertheless, since we know that the Sun itself is in a state of continuous and variable agitation, it is natural that we should ask ourselves whether this variability may have an influence on the climate of the Earth.

In order to be able to answer this question it is necessary, among other things, to measure the energy which reaches us from the Sun when it is quiet, that is when it is not affected by sunspots or eruptions, and compare it with the energy emitted by the disturbed Sun. The comparison will show whether the energy varies during the 11-years' cycle or follows any other cycle of different period.

This type of investigation is not an easy one. The radiation from the Sun, before reaching the surface of the Earth, must travel through our atmosphere which varies in density and transparency in different parts of the Earth according to seasons and at various heights. Great progress, however, has been made in this field in recent years. By choosing selected stations in places where cloudiness is at a minimum (Chile, South Africa and Israel) it has been possible to show, with reasonable accuracy, how the 'solar constant' is actually slightly variable with time. Astronomers and physicists define the 'solar constant' as the quantity of heat received per minute by a square centimetre of the surface of the Earth. It is called 'constant' because observations carried out until a few years ago indicated that this quantity did not change. Nowadays the more delicate instruments available and the selection of suitable sites where the atmosphere is pure, seem to indicate that this quantity of heat is actually variable from season to season and from year to year, even if the variation is very small.

It has not been possible, so far, to establish a direct and definite correlation between the 11-years' cycle of the Sun and the variations detected in the solar constant. The very fact that there is an indication of variation in the solar constant leads to the consequence that on the Earth there must be some unevenness in the normal heating and cooling of its various parts. As a result of this, movements of large masses of warm and cold air must follow which, in their turn, may be responsible for meteoro-

110

logical perturbations in different regions of the Earth. The heating of the air masses changes continually because of their very complicated movements produced by the rotation of the Earth around its axis and on account of the inclination of the equator with reference to the ecliptic. It is not surprising, therefore, that meteorologists find it extremely difficult to establish how the accidental variations of the solar constant may influence this or that region of the Earth thereby producing droughts or heavy rains or various amounts of clouds. On the Earth there occur years of good or bad weather as farmers know only too well, and it may be thought that such periods are due to solar influence. We are still very far from being able to establish how this happens and even further from being able to predict meteorological conditions based on the variability of the solar activity, although of course, it is comparatively easy to predict in general the actual activity of the Sun.

We can, however, define the influence of solar energy on terrestrial vegetation, and perhaps in the near future we shall be able to make use of this knowledge following the development of scientific experiments and investigations. On the one hand, the astronomer and the physicist can establish which are the properties and characteristics of solar radiation and its variations, while on the other, the farmer, by making use of this information, may be able to increase the soil production.

The use of solar radiation has given rise to very important studies which are directed towards its practical application. The solar radiation is known to us mainly as white light whose intensity depends on the inclination of the solar rays which reach the surface of the Earth. When we analyse this white light by means of glass prisms or by other methods we discover that the light is made up of radiations of all colours, from violet to red. Beyond the violet and the red there are also other radiations to which our eyes are not sensitive and which we call ultraviolet and infrared. Moreover we discover that our atmosphere filters the solar radiation and stops part of it, particularly in the ultraviolet, so that at various heights above sea level the total radiation received from the Sun is different because the depth of the atmosphere interposed between us and the Sun is less.

What influence have these various types of radiation on the growth and development of plants? Light acts upon plants in such a way that they can extract the carbon from the carbon dioxide present in the air and which penetrates the leaves through the many tiny holes with which they are provided. In the cells of the leaves, carbon combines with water and with the salts which are absorbed from the ground through the root system, and produces very complex and vital substances such as carbohydrates and proteins. This process which goes under the name of photosynthesis, is the basis of life on the Earth because plants feed animals and both plants and animals feed man. The growth of plants requires certain conditions of light and temperature and both these are supplied by the Sun.

Visitors to the Observatory of Juvisy, near Paris, which was founded by Flammarion, the well-known author of popular texts on astronomy, may discover a derelict greenhouse with compartments covered by glasses of various colours. This greenhouse was used for experiments devised and carried out by Flammarion in order to study the action of solar radiation upon the development of vegetation. Since Flammarion's time these experiments have been continued, though in a much more refined form, but we are still only at the beginning of the investigations in this branch of the physiology of plants.

The function of chlorophyll, which is extremely important in the process of photosynthesis, depends essentially upon the energy supplied by the red radiation. On the other hand the violet radiation appears to play an important part in the phenomena of synthesis on which the metabolism of plants largely depends. Experiments have been carried out by illuminating seedlings of wheat or oats with lights of different colours but of the same intensity. It seems that the whole visible spectrum, namely from red to violet, is efficient in the production of photosynthesis. A group of radiations in the extreme red and another in the blue, appear to have a more energetic action and of the two that in the red is more intense. Perhaps the action of the different radiations may be different for the various species of plants and a good deal of ground is still to be covered in these experiments before these questions can be completely understood. Once these problems are solved it will be possible to make better use of the solar radiation, both in different regions of the Earth and at various heights above sea level. Experimental stations may be built with particular attention being paid to their position, to the selection of suitable solar radiation, to the distribution of temperature, of humidity, and so on.

We have already mentioned that the periodic variation of solar activity may have an influence on the climate of the Earth. It has been suggested that the behaviour of vegetation, and in particular the growth of wood in trees which generally have a long life, may in some way reflect the variations of solar energy due to the phenomena which occur on the surface of the Sun. A study has been made of the rings of growth of wood in *Sequoia Gigantea* in California and in Arizona, where the climatic conditions are more stable and where at the same time humidity and precipitation, which are essential conditions for the life of plants, are more regular. The rings of growth studied in sections of the trunks of these trees show some groupings which are very close and others more widely spaced. The number of rings in each group is about 11, which agrees with the number of years of the solar cycle. Indeed, in these ring systems a grouping of 22 years is more noticeable, thus corresponding to two solar cycles. As we have already mentioned earlier, this represents better a complete solar cycle if we take into account the magnetic characteristics of sunspots. Some of the trees studied were nearly 2,000 years old, and they showed very clearly their rings of growth which indicate the variation of growth that these trees undergo in relation to the maximum of the cycles of solar activity during their whole life. In addition they also indicate periods of greater or lesser precipitation of the regions where these sequoia grew.

Vercelli has studied the trunk of one of these sequoia whose life spanned the period from 274 B.C. to A.D. 1914, namely nearly 2,200 years. He detected very clearly a cycle of 11 years whose maxima and minima are in phase with the actual frequency of sunspots, at least for the period for which we have records of sunspots. Evidence of this particular phenomenon has also been found in other trees, though not as ancient, in the German forests and in the pine-woods of Ravenna.

Temperature, pressure, rain and relative humidity are meteorological factors of extreme importance for the growth of plants, and we know only too well how such factors can be variable and different even in regions which are close to each other. Before we can clarify the cause of the phenomenon, we ought to know what are, generally, the average climatic conditions of the Earth as a whole. We ought also to study the variations corresponding to the fluctuations of the ultraviolet and corpuscular radiations which reach the Earth with varying intensity during the solar seasons. We are faced, therefore, with the study of huge phenomena over vast regions of the Earth. For instance, it has been discovered that the average level of the great African lakes (Nyasa, Victoria, Albert), of the Caspian Sea and of other large basins, is noticeably higher during the maxima than during the minima of solar activity. This can be taken as an obvious indication that in these regions, rain must have been more intense during the maximum of solar activity.

There is also an indication that the flooding of the Nile follows the solar cycle in the sense that the floods are more extensive during the years when the Sun is active, leading us to the conclusion that generally speaking, rain amount is much greater, on the Earth as a whole, during these periods. Another phenomenon which has been noted is that at the same time a much greater number of thunderstorms are recorded in the tropical regions.

During the maximum of activity of the Sun there is some evidence that the general circulation of the terrestrial atmosphere is more agitated and this could well produce a general temperature on the whole of the Earth a little lower than the average. This may be true for the tropical regions but at temperate latitudes, the phenomenon is rather irregular.

In the case of the growth of plants it is thought that the general meteorological conditions are not the only ones to have an effect, but that another important factor is the quantity of ultraviolet radiation which is emitted by the Sun. We already know that a large amount of this is absorbed by the layer of ozone at about 15 miles in the atmosphere of the Earth. Nevertheless, when we climb to the top of mountains, at heights much less than this, we soon detect the influence of such radiation on our skin, an influence which increases rapidly as we climb higher above sea level. During periods of violent activity of the Sun, such as occur during the maximum of its activity, large amounts of ultraviolet radiation, of corpuscles, of atoms, of electrons and of cosmic rays are emitted by flares. All these forms of radiation which reach the Earth, but more especially the ultraviolet radiation, may be the cause, more or less directly, of a more vigorous growth of plants, provided that no other factors intervene which may act in opposition.

Another class of terrestrial phenomena which is directly linked with solar phenomena, is the display of aurorae which are manifestations of an electromagnetic nature and which appear in the higher layers of our atmosphere. Since 1850 a correlation has been discovered between the 11-years' cycle of the Sun and the variability of terrestrial magnetism as is daily determined by the geophysical observatories scattered over the Earth. In addition to the regular and periodic variations which follow the general activity of the Sun, others are observed which have the characteristics of a violent perturbation called a 'magnetic storm' and which, very often, show a close correlation with solar storms. Aurorae, which are generally very intense, are seen with great frequency between the Arctic and Antarctic circles and the poles, and show a very marked link with terrestrial magnetism.

All the polar explorers have given detailed descriptions of aurorae. In Norway and Sweden special observatories have been established to record, in detail, such events and to photograph them. Since the connection between aurorae and the phenomena of terrestrial magnetism, and hence solar phenomena, has been established, it is obvious that the origin of aurorae must be due to an emission by the Sun of electrically charged particles. Such

115

radiation is deflected towards the polar regions by the terrestrial magnetic field and tends to concentrate there. The gases of which the stratosphere is composed, are subject to excitation and, as a result of this, the upper atmosphere, in places, becomes bright with patches of brightness which move very rapidly, probably on account of the movements of the showers of corpuscles which reach us from the Sun. These particles must leave the Sun in large numbers, like a beam with a very large angle so that, even if the disturbances producing them are near the limb of the Sun, that is to say away from the line joining the centre of the Earth and the centre of the Sun, they can still envelope the Earth.

In the region of the North Pole, the zone of maximum frequency of aurorae passes through the North Cape, Labrador, the southern part of Greenland and Iceland. This zone contains both the geographical and magnetic poles, and something similar occurs in a corresponding zone of the southern hemisphere. At lower latitudes the phenomenon is more rare. In order that aurorae might be seen even in southern Europe, the activity of the Sun has to be exceptional. This happened during the maximum of solar activity in 1870 and it has also happened in more recent maxima.

It may be of interest here to recall the facts concerning the great aurora observed over a large part of the Earth on the night between the 4th and 5th of February 1872, which can be compared with the aurora observed over most of Europe on the night between the 25th and 26th of January 1938. Donati, who was the director of the Florence observatory and later became the director of the Arcetri observatory, asserted that the aurora was due to solar causes. In 1872 the Sun had already passed its maximum of activity which occurred in 1870 and which, for its intensity, had been a very remarkable maximum. The aurora of 1872 excited so much interest that Donati sent a circular letter to all the Italian consuls abroad in order to obtain as much information as possible. He later included the information he had collected in a memoir which was to be one of his last published works. Having compared the times when the aurora appeared in various places, he concluded thus: 'The luminous phenomena of the great aurora which was observed over an extremely large part of the Earth on the night between February 4th and 5th, 1872, were first seen in the east and then in the west and showed themselves in the various parts of the Earth almost at the same local time, with a slight tendency to gain on this time as the phenomenon moved from east to west.' Having in this way established that the apparition of the aurora was linked with the apparent path of the Sun and hence could not be attributed to a phenomenon of terrestrial meteorology, he concluded that it was necessary to think in terms of a new 'cosmic meteorology'. 'The phenomenon of sunspots' he continued, 'and all the other phenomena which occur on the Sun, are in fact outside the domaine of the old meteorology, and if the aurorae are related to phenomena which take place on the Sun, then we are led to think that they are not produced by meteorological causes only, but by cosmic causes as well.'

The newspapers of the time gave ample coverage to the exceptional aurora of the night of January 25th, 1938. The phenomenon lasted from sunset until morning with varying intensity and, naturally, was much more conspicuous in the northern parts of Europe, particularly in the neighbourhood of the Arctic circle. At the observatory of Pino Torinese, at 21.00 hours the whole northern sky was coloured by a deep pink light through which the stars were clearly visible. The presence of clouds enhanced the beauty of the display. The long chain of the Alps, from Rocciamelone as far as and beyond Superga, seemed to be hung with red drapery which extended over a very large area of the sky near the constellations of Cassiopeia, Cepheus, Ursa Major and Ursa Minor, reaching the zenith almost in the shape of a triangle. On the side of Cassiopeia and in the neighbourhood of a cloud, all of a sudden a beam of light appeared of a colour between white and yellow, which stretched for several degrees towards the zenith, resembling the beam of a searchlight. The phenomenon gradually decreased in intensity, but towards one o'clock in the morning the red light of the aurora appeared again in all its splendour.

At Bologna the various luminous rays of the aurora were seen stretching towards the zenith and moving in the sky from east to west. Up to seven were counted and they were white or pale pink in colour. After 22.00 hours they disappeared and only a diffuse pale light was left towards the north. On the following nights, particularly in northern Europe, other aurorae were seen but none as intense as the one we described.

Observations of the Sun made with the solar tower at Arcetri showed a large group of sunspots which crossed the central meridian on January 18th. This group was observed continually until it set at the western limb of the Sun on January 24th. At the point of the limb where the group set, several prominences were photographed which appeared to consist of long filaments which, like an enormous fountain, erupted from the solar limb ascending rapidly into space. It is very probable that this eruption was the cause of the aurora which was seen on Earth on January 25th.

A similar phenomenon occurred on Easter Day, March 24th, 1940. On the morning of this particular day, clouds prevented the usual photographic observations of the Sun in hydrogen light (Hα) with the solar tower at Arcetri. These would have given us a very definite indication of the conditions of the solar atmosphere. However, only visual observations were possible on that day. The visual observations revealed that a rather large group of spots, consisting of several nuclei, which had appeared on the east limb of the Sun on March 20th at a latitude of 12° north, was surrounded by an unusually dark and well developed penumbra and with indications of very strong activity. The Earth responded almost at once to this phenomenon. From 19.00 hours to 19.30 of that day, a beautiful aurora was visible in Italy. In Florence, although the sky was not completely dark because of twilight and of the presence of the rising Moon, we were able to observe the aurora which consisted of two large red clouds, occasionally changing their colour into yellowish-white. As well as the colour, the shape and position also changed, moving between the constellation of Draco in the north-north-east to that of Cepheus in the north-north-west. The clouds with their drapery and arc effects, sent out long rays which almost reached the zenith. The intensity of the red light and of the aurora clouds was at first so great as to attract the attention of many passers-by in the street. From Arcetri, towards Fiesole and Mount Morello, the display was really fantastic. After 20.00 hours nothing was left in the sky, which was becoming brighter because of the presence of the Moon, except a long reddish cloud in the north-west.

The very same day when these phenomena occurred, a magnetic storm of exceptional violence was unleashed over the whole of the Earth. In the recordings obtained by the magnetometers of the Hydrographic Institute of the Navy at Genoa, there is clear evidence of a strong disturbance in all the components of the terrestrial magnetic field at 14.45 hours. At the same time the ionosphere was also so disturbed that shortwave radio communications on the Earth became impossible.

From what we have said and from the fact of the existence of a strong correlation between solar disturbances and terrestrial effects, it may be thought that whenever a spot appears on the Sun, we should immediately expect some response on the Earth. Actually this is not so because the phenomenon is very complex and groups of spots, sometimes even large groups, have been known to appear on the Sun without any effects, or at least any large effects, being detected on the Earth. More than the solar storm itself it is its intensity which counts and its development which often depends on and is very much influenced by the presence of solar flares. In other words we cannot claim that we have discovered a precise correlation between solar and terrestrial phenomena. It is true that many observations have given us the certainty that such a relationship exists, but many more observations, both of the Sun and of the Earth, will be required before we can clarify to the full, and in every detail, the whole of these imposing phenomena.

Part II EARTH, PLANETS AND COMETS

1 · The Journey of the Earth in Space

It is probable that many of the engineers who are involved in the design and construction of missiles, rockets and satellites of increasing power must have often thought that nature has already provided us with a wonderful and comfortable spacecraft which can hardly be improved. After all, this is what our Earth is. For millions and millions of years it has been travelling through cosmic space and probably it will continue to do so for a time equally long. In its complex and fast motion it carries many natural wonders which exist on its surface and, thanks to the benevolent influence of the Sun, ruler of the solar system, it enables animal and vegetable life to exist.

Spacecraft built by man have the advantage of not being restricted to a predetermined course, but can travel to any particular destination in space. Nevertheless, this choice of destination is limited both by human technology and physiology. Often we hear, even at international conference level, talks of cosmic space as if it had already been conquered by man. Terrestrial regions of difficult access, such as the Arctic and the Antarctic, have been conquered, but man seems to forget the immensity of the universe. There is also another problem which will probably remain beyond the solution of man, and that is how to approach our Sun, much less the large suns which populate the universe, once we have left our own Earth.

Let us come back to our 'spacecraft' and its complicated journey through cosmic space. What is our present knowledge and what can we see from our terrestrial observatory? Everything on the surface of the Earth, both animate and inanimate, takes part in its diurnal rotation, completing one rotation around the axis of the Earth in an interval of time of 24 hours and moving in a direction opposite to the apparent motion of the Sun in the sky. At the equator we move from west to east at a speed of 1,525 feet per second, and this speed decreases as we approach the poles. In this way we can witness during our diurnal journey, the wonderful spectacle of the rising and setting of the Sun, of the succession of twilight, of day and of night and of the appearance of the stars in the sky, first the brightest, followed gradually by those less bright. We witness the appearance of the luminous Milky Way, of the planets and of the Moon which travel among the fixed stars. The spectacle increases in interest when we realize that the whole canopy of the heavens seems to move from east to west, as a consequence of the rotation of the Earth, faster near the celestial equator and slower decreasing to zero when we look towards the pole, marked by the Pole Star in the northern hemisphere, and in the southern hemisphere by a point in the neighbourhood of the Southern Cross.

This is only one of the motions of our natural 'spacecraft'. There are, however, many others much faster and of longer duration. Let us take the motion of the Earth around the Sun, which is accomplished in approximately 365 days. This particular journey offers to us various wonderful and interesting sights, even if they tend to repeat themselves every 12 months. Those who are interested in the night sky, can easily see how stars and the constellations they form rise and set every evening a little earlier with reference to the Sun. Slowly, in this way, the night sky changes its appearance in the course of the seasons.

The Sun appears to move among the stars, but in actual fact it is the Earth that in describing its annual orbit around the Sun, enables us to see the whole of the sky. In our journey around the Sun we shall, therefore, see those constellations during the winter which are not visible in the summer, because then they are in the sky in day time. Our 'spacecraft', therefore, in addition to its rotation around its axis, also revolves around the

Sun with a velocity of about 18 miles per second. Man has been able to determine this velocity with great accuracy, so that now we have no doubts about the reality of this motion which for centuries proved to be so difficult to understand. We could call this second motion around the Sun, the planetary journey of our 'spacecraft' which is compelled to follow always the same path. The result is perhaps a journey without great variety of view, but nevertheless majestic and accompanied by many interesting events.

Further surprises are in store for us. No celestial body is at rest, all are in motion in cosmic space and subject to mysterious gravitational forces; the larger the mass the greater the force which correspondingly increases for the total mass of the bodies which populate the sky. Thus the Sun with its cortège of planets cannot be stationary in space. Only in recent times have we been able to understand to what force the Sun is subject, a force much greater than that which the Sun itself exerts upon the planets.

The Milky Way which we can see in the sky as a whitish band and which consists, in reality, of millions of stars and of cosmic matter, has a centre. For the time being we know of the existence of this centre but very little more. Around it and at various distances from it, move stars and cosmic matter more or less as the planets move around the Sun. The Sun itself is rather far away from the centre of the Galaxy but it revolves around it accompanied by the whole of the solar system. Observations and calculations reveal that the Sun revolves around the centre of the Galaxy with a velocity of about 174 miles per second, and completes one revolution in the course of 200 million years.

It is wonderful to realize how well we are enveloped and protected by the atmosphere which surrounds our natural spacecraft, so that we do not feel any of these many motions. Since we are carried around the centre of the Galaxy on a path so much longer through cosmic space than the relatively short path around the Sun, we ought to be able to explore new regions through which the solar system travels. Actually this does not happen. Why is it that in our extremely long journey around the centre of the Galaxy, the general view does not change? Why is it that we do not see new stars or we do not approach other bodies close enough to study their physical characteristics in detail as we can do in the case of the Sun?

The explanation is very simple. Let us suppose that we go through a wood with trees widely scattered and very distant from each other as well as from our path. We would notice a similar effect to that which strikes us when travelling by train or by car. The trees nearest to us appear to move much faster than those further away, while those very far away indeed, appear to be practically stationary. Let us now extend this effect to the dimensions of the Galaxy, which, although very large to us, is but a minute object if compared to the universe as a whole. It is clear, then, that considering the distances between the stars which form the Galaxy, even at the velocity of 174 miles per second at which we travel, thousands and thousands of years will be needed before any variation may be seen in the configuration of the stars as observed from the Earth which would show that the solar system is moving among them.

It is indeed impossible to predict the destiny of humanity and of our 'spacecraft' in the course of the next few thousand years. One thing is certain however, our 'spacecraft' continues its interstellar journey, undisturbed and only the brevity of human life prevents us from admiring any other panoramas and wonders in the immensity of the Creation.

2 · The 'Tail' of the Earth

One of the most spectacular wonders among those which nature offers us is that of the large atmospheric envelope which surrounds the Earth and accompanies it faithfully in both its diurnal rotation around its axis and its revolution around the Sun. A question which comes to mind is whether this envelope is distributed uniformly around the Earth or is subject to variations particularly when it is illuminated by the Sun. The atmosphere is relatively dense near the surface of the Earth and becomes increasingly more tenuous further away from the surface and has hence an undetermined limit in interplanetary space.

The force of attraction that the Earth exerts upon all bodies is also responsible for holding down the atmosphere. This is true only for the heavier gases of the atmosphere such as oxygen and nitrogen, but the lighter ones, such as hydrogen and helium, are able to escape into space. Thus, although these gases are continuously produced by various terrestrial sources, they are rather rare in the atmosphere. On the other hand, we have already discussed the fact that the Sun is surrounded by a very extensive envelope, the corona, which is clearly visible during a total eclipse of the Sun. Photographs of the corona taken by observers in aeroplanes high above the Earth, but not yet by rockets or satellites from outside the terrestrial atmosphere, have revealed that the corona extends much further than it appears from photographs taken on the surface of the Earth. Indeed the corona extends so far as to suggest that it may well extend in space as to merge with the zodiacal light. This is another wonderful phenomenon which can be best observed on the Earth at medium latitudes or even better in the tropical regions.

In these regions, immediately after sunset, or just before dawn, when the sky is still fairly dark, the zodiacal light can be seen. The name is due to the fact that this light extends in that band of the sky known as the zodiac, where the Sun is to be found throughout the year. From that point in the sky where the Sun has set, or is about to rise, we can see rising along the zodiac a luminous band which is brighter nearer the horizon. This band becomes fainter and narrower the further it reaches from the horizon until it ends at the zenith in a point. As the zodiacal light sinks below the horizon, on account of the rotation of the Earth, a weak light appears, if the sky is very clear, at a point almost exactly opposite to the Sun. This light is known as 'gegenschein' which translated literally means 'opposite light'.

If it is true that the solar corona extends so far as to merge with the zodiacal light, then we can conclude that the Earth is enveloped by an extensive cloud which is lens shaped, with the Sun at its centre, and which extends along the solar equator beyond the orbit of the Earth. The composition of both the corona and of the zodiacal light, has been studied by means of spectroscopes. Nowadays our knowledge has been considerably increased by the possibility of sending spectroscopes beyond the terrestrial atmosphere and by means of radio telescopes. The gas of which the corona is composed consists mainly of protons and electrons and of metallic atoms, such as iron, calcium and nickel, which are excited and at a very high temperature.

The zodiacal light appears to be due to the reflection of sunlight by millions of extremely minute meteoric particles, which probably are of the same type as those which form Saturn's rings, mixed with hydrogen and possibly other gases as well. We know that periodically on the surface of the Sun, that is in the photosphere and in the chromosphere, below the corona, large storms take place which emit gases. These gases, among which hydrogen predominates, travel through the corona and have enough energy to leave the sphere of attraction of the Sun, and in due course they reach the Earth in the form of ultraviolet radiation, or as atoms or as parts of atoms. A very active exchange, therefore, takes place between the solar and terrestrial atmospheres. The former impinges upon the latter with powerful radiations and emissions of matter, while from the terrestrial atmosphere the lighter gases are dispersed into space. In this way, probably, is established an equilibrium which allows the terrestrial atmosphere to maintain its characteristics so necessary for the upkeep of our animal and vegetable life.

By means of modern methods of research we have been able to establish that the distribution of hydrogen around the Earth, as far as approximately 10 terrestrial radii, is almost spherical. This forms an envelope which can be called 'geocoma', namely 'terrestrial coma'. From the geocoma is emitted the greatest part of the diffuse ultraviolet radiation which can only be observed when we go beyond the terrestrial atmosphere, since this tends to absorb it. It has also been discovered that on the side of the Earth not exposed to the Sun, the hydrogen from the geocoma extends into a tail. We have here a similarity with comets which generally have tails which always point in a direction opposite to that of the Sun. The tenuous extension of the hydrogen cloud which surrounds the Earth must be produced by pressure of sunlight which repels the light atoms of the gas by overcoming the force of attraction.

That the Earth is followed by a tail composed of hydrogen and possibly of other gases, directed always away from the Sun, is not a new hypothesis. It was formulated nearly half a century ago following observations of the gegenschein. Now, after new investigations, it is possible to confirm that this light originates in the tail which extends away from the Earth as far as 1,000 terrestrial radii and which has a width of 100 Earth radii. Probably in the tail, the hydrogen atoms and the lighter corpuscles are repelled with a greater force than the heavier atoms such as those of helium. Because of the revolution of the Earth around the Sun, the tail must appear curved in space, just as we see curved tails in the case of comets.

If this theory is correct, future astronauts when they travel far enough into space, will be able to see the Earth illuminated by the Sun, followed by a long tail, just like the one observed in the case of Halley's comet at its last apparition in 1910. Perhaps the astronauts will also be able to observe the tails which are thought to follow the other planets. Unfortunately the light of these tails is too faint, and is invisible from the Earth because it is absorbed by the terrestrial atmosphere.

3 · Aurorae — The Light of the Night Sky— Noctilucent Clouds

We owe most of our knowledge about aurorae to Karl Stoermer, the Norwegian scientist who was a pioneer in this work.

The awe-inspiring spectacle of the aurora, visible in its maximum splendour in proximity of the magnetic poles of the Earth, has been described, painted and photographed (Plate 13) by polar explorers as well as by observers stationed in suitable places. Many stations specializing in this work have scientifically studied aurorae in their various aspects. Scientists have paid particular attention to their periodicity, their frequency at various latitudes, the heights at which they occur and their physical constitution.

The aurora is known as 'aurora borealis' if it occurs in the northern hemisphere and 'aurora australis' if seen in the southern hemisphere. Since the aurora borealis appears in regions and continents which are inhabited there is naturally more data of observations available about it than about the aurora australis.

Aurorae were well known by the ancient world and we find descriptions of them both in Greek and Latin authors. Seneca in his *Naturales Quaestiones* describes the phenomenon in the following terms: '. . . sometimes flames either stationary or in motion can be seen in the sky. They are of various forms, some are like a luminous corona within which the celestial fire is missing. They thus form something like the entrance of a cave. Others have the shape of a great luminous barrel which either moves from one point to another or is stationary. Others again

131

have the appearance of large bays emitting flames, which at first seem to be hidden in their depths. These fires are of various colours. Some are of a bright red, others are like a weak flame, others are white, some sparkle and yet others are of a uniform yellow. Historians have often recorded these phenomena. At times these fires are so high as to shine among the stars or are so low that they look like the reflection of a distant fire. This happened at the time of Tiberius when the cohorts raced to Ostia believing that it was on fire. During most of the night the sky was illuminated by a weak light which looked very much like a dense smoke.'

The great aurorae of the years 1870 and 1872 were also observed in Italy, as we have already stated (page 116) and were the object of interesting investigations carried out by the early Italian spectroscopists.

Weyprecht, who in 1872–1873 spent a winter with an Arctic expedition in Franz Josef Land, gives us a very vivid description of the aurora: '. . . and here is a new form. Bands of every possible appearance and intensity and shafts of light move in the sky. Towards the south, a weak band which is hardly visible, extends along the horizon; suddenly it seems to lift rapidly and to extend both towards the east and the west. The luminous waves cross each other and some rays rise up to the zenith. For a short time they appear to stop and then they come to life once again. The bright waves oscillate from east to west, their edges take on dark red and green colours and dance up and down. The rays suddenly jerk rapidly upwards and become shorter: then all rise together and tend to approach closer and closer to the magnetic pole. It looks almost as if a race among the rays is in progress, as if each one is bent on being the first to reach the pole. Once the pole is reached, the rays spring towards all the points of the compass. Do the rays strike from above to below or vice versa? Who can tell? From the centre a sea of flames seems to spread: is this sea red, white or green? It seems to be of all three colours at the same time. The rays reach down to almost the whole of the horizon and the whole sky is on fire. Nature offers such a display of fireworks which exceeds human imagination. Almost against our will we are listening, in fact we feel that such a spectacle must be accompanied by some sound. Instead a deep silence reigns all around us: not the slightest sound reaches our ears. Suddenly the whole phenomenon disappears, with the same speed with which it appeared and the darkness of the night again envelopes us.'

Nowadays photographs in black and white, coloured photographs and cine films help the imagination of those who have never witnessed such a phenomenon. They also help to classify the various forms of aurorae in pulsating bands and homogeneous arcs; in drapery, in rays and in the form of corona.

By observing aurorae at various latitudes it has been possible to trace a map of the distribution of aurorae in the northern hemisphere in which the curves known as 'isochasms' show the places which have the same auroral frequency. From these maps it is clear that aurorae have the magnetic north pole for geometric centre. The magnetic north pole is situated on the western coast of Greenland. If we compare places in Europe with the same latitudes as those in the United States we find that aurorae are more rarely seen in Italy, for example, than in the state of New York. While 15 aurorae can be observed in one year in New York, in Rome only one aurora can be seen in 10 years. In north Canada, in Norway and along the Arctic coast of Siberia, 243 aurorae can be observed. It is therefore natural that in these regions of the world special stations have been set up for the purpose of observing aurorae, and all these stations co-operate in an international programme.

Halley, in 1716, noted that aurorae generally appeared when a perturbation of the terrestrial magnetic field occurred. These perturbations, which we call 'magnetic storms', reveal themselves by variations of the characteristics of terrestrial magnetism which can be measured by special instruments.

The Earth behaves like a huge magnet, surrounded by a well-known system of lines of force, which determines the terrestrial magnetic field. This field shows variations which are regular and irregular. The regular variations are dependent upon the diurnal and annual position of the Sun with respect to the Earth, while the irregular variations depend on the activity and on the disturbances which take place on the Sun. In the irregular variations of the terrestrial magnetism there has been detected the presence of an 11-years' cycle which coincides with the 11-years' cycle of solar activity. Since the formation of aurorae is also strictly linked to the perturbations of the terrestrial magnetic field, it is obvious that both the presence of aurorae and the existence of magnetic storms, must be the result of solar storms. It is the Sun, therefore, which is the cosmic cause of these phenomena, as it is for so many other terrestrial phenomena (see page 115).

It is of great importance to determine the height at which the aurorae are formed as well as which of the atmospheric layers are responsible for their formation. Special studies have been carried out in Norway, in Canada and in the United States, in particular at Cornell University where the National Geographic Society, in 1938, set up a centre. The investigations consisted in the determination of the heights of the various types of aurorae, by obtaining simultaneous photographs at various stations separated by a known distance. By means of this system of triangulation and by reference to stars which are visible in the sky at the time of the aurorae, it has been an easy task to establish their height. No aurora has been observed at a height of less than about 22 miles above the surface of the Earth. The highest, measured by Stoermer, was at a height of 621 miles. The majority of aurorae occur between 56 and 75 miles and homogeneous arcs rarely exceed 93 miles.

In order to record the changing form and the development of aurorae, Gartlein, at Cornell University, employs an automatic camera which can take a photograph every thirty seconds. In this way he has been able to obtain a very large number of photographs of aurorae. From these investigations it is possible to conclude that aurorae are formed above the stratosphere, in the ionosphere, extending through the E and F layers which reflect radio waves. This is a very important discovery which enables us to understand how aurorae are formed. Extensive research is being carried out in order to determine the quality of the light emitted by aurorae. Spectral analysis, as in the case of the Sun and of the stars, has solved this problem. The measurement and interpretation of the spectrum of the aurora has contributed considerably to remarkable discoveries both in the field of physics and geophysics.

Gauss in 1838 had already suspected that the main agent producing the aurora was electricity in motion. Ångström was the first to observe the spectrum of an aurora. By pointing an ordinary spectroscope towards the brightest parts of the aurora, he noted the absence of a continuous spectrum and the presence instead of a very bright and intense yellowish-green line, three blue lines and a red line. If we observe with a spectroscope the light of any coloured light tubes which are found nowadays in streets, shops and illuminated advertisements, and which contain rarefied gases of various elements excited by an electric discharge, we see similar spectra, although generally, the lines will be of different intensity and occupy a different position in the spectrum. While using spectroscopes of limited power it was not easy to identify the elements which emitted the lines in the spectrum of the aurora. The difficulty arose because the extension of the spectrum between violet and red was not large enough and therefore it was not possible to determine accurately the position of the lines, or in other words their wavelength.

In the last 30 years, photographs of the spectrum of aurorae have been obtained. This was made possible by the progress in the design of instruments and in the production of more sensitive photographic emulsions, even if in some cases the exposures had to be very long. On the spectrograms obtained, about 40 emission lines and a few bands were measured. Spectroscopists have established that the lines are due to atoms of the elements in a state of excitation which is either thermal or produced by electric discharge. The bands, on the other hand, so called because of their appearance, are produced by molecules, namely by combinations of atoms which have been somehow excited. The first bands which were identified were those due to nitrogen, which is a well-known component of the atmosphere. Molecular aggregates of nitrogen atoms are hit by electrons in the ionosphere and as a result they emit bands in the violet and in the blue regions of the spectrum. As we have already mentioned, the most intense line of the spectrum of the aurora is of a yellowish-green colour. For a very long time this presented the spectroscopists with an insoluble problem. At last, however, the line was identified as being produced by an atom of oxygen, another well-known component of our atmosphere. The wavelength was at first accurately obtained by interference methods. Later McLennan and his associates at Toronto University were able to obtain the green line of the aurora by electric excitation of oxygen atoms. The reason which made the identification of this line so difficult is to be found in the actual conditions of the electric discharge to which the oxygen atom has to be subjected in order to produce the line. A discharge in a vacuum tube containing pure oxygen, produces the line in question although very weak and in addition masked by the band of molecular oxygen. If, however, a rare gas, preferably argon, is mixed with oxygen, then the green line is much more enhanced with reference to the rest of the oxygen spectrum. Probably the reason for this is due to the fact that the rare gas has the effect of limiting the spectral lines of the oxygen to those only of low excitation such as the green line. This discovery led to the prediction of the existence of three lines of oxygen in the red region of the spectrum of the aurora. The lines were duly observed in the spectrum of the aurora and in that produced in the laboratory. Other lines are due to ionized atoms of oxygen and nitrogen, that is to atoms which have lost one or more electrons on account of high excitation. The other lines of the auroral spectrum have not yet been identified.

It is interesting to call attention here to the fact that in other very rarefied atmospheres, such as those of nebulae and of the solar corona, we find very intense emission lines in the green, as well as in other regions of the spectrum. In the case of nebulae the lines are also produced by atoms of nitrogen and oxygen which are in such a high state of excitation that they have not yet been reproduced in terrestrial laboratories. Even higher must be the excitation of the atoms of metals which emit the lines in the spectrum of the corona and which are so difficult to interpret.

It was suggested at first that the causes producing aurorae were to be found in the Sun. Investigations showed that in actual fact aurorae are due to enormous swarms consisting of electrons, atoms, ultraviolet radiation and of corpuscles which are propelled beyond the gravitational field of the Sun and travel for millions of miles in space and finally may even reach our small planet. Generally speaking these swarms are emitted by the Sun as a whole, that is to say by its photosphere, but only when the latter is affected by disturbances such as storms which may occur during the 11-years' cycle. We can imagine these swarms as being emitted from regions where sunspots are present, or from disturbed regions in their neighbourhood, or from other regions which, although without spots, are nevertheless disturbed. At times they may originate in regions that appear to be undisturbed and are emitted in forms which look like a jet of water ejected under pressure from a hydrant. Since the Sun rotates around its own axis, every time these swarms collide partially or totally with the Earth, they produce on it and in its atmosphere various phenomena among which we may include the aurorae.

Stoermer has shown both theoretically and by calculations, that as the ultraviolet radiation and the swarms of atoms and corpuscles approach the Earth, they are subject to the influence of the terrestrial magnetic field. As a result they will tend to show a preference to crowd around the magnetic poles and collide with the atoms of the extremely rarefied air present in the ionosphere. Once these atoms are excited they emit light which we admire in the various shapes of the aurora. In the observation of aurorae we should expect a repetition of the phenomena every 27 days, which is the period of rotation of the Sun as seen from the Earth. This periodicity actually occurs and we can therefore expect the return of aurorae when the disturbed region of the Sun next appears and we should also expect a frequency period of 11 years.

Until it became possible to determine whether the appearance of an aurora was simultaneous with the appearance of a disturbance on the Sun, or whether there was a time lag between the two phenomena, observers could not decide whether the aurorae were related either to the swarm of corpuscles or the emission of ultraviolet light from the Sun. Observations, however, showed that in general, aurorae develop about 26 hours after the appearance of a solar disturbance, and therefore they must be produced by the swarm of corpuscles which must enter the atmosphere of the Earth on that part which is not yet turning towards darkness. The geographical distribution of aurorae with reference to the magnetic field of the Earth, shows that the corpuscles cannot be neutral but must be electrically charged. Were the corpuscles neutral they would penetrate the terrestrial magnetic field at any latitude and aurorae would be more frequent than they are in the equatorial regions. At the same time as an aurora is observed, magnetic storms are recorded on the Earth and this fact is an indication of the arrival of the swarm of corpuscles in our atmosphere and of the ionization that they produce.

Although we can admit that considerable progress has been achieved in the observation and explanation of solar phenomena in relation to terrestrial phenomena, there is still a considerable amount of work to be carried out by theoretical investigators. They have the problem of explaining well-defined terrestrial phenomena which are produced by causes originating in the Sun. Both the origin of these causes and the way they are transmitted towards the Earth are still a mystery. Mathematicians and physicists are working to solve these very complicated problems of electrical, magnetic and in part hydrodynamic nature. Meanwhile the layman can enjoy and admire the awe-inspiring and beautiful phenomenon of the aurora.

The zodiacal light and the aurorae are not, however, the only phenomena, apart from the celestial bodies, which give light to the night sky. There are two more phenomena known to geophysicists as the 'light of the night sky' and the 'noctilucent clouds' which, although not as conspicuous as the aurorae are nevertheless of great interest.

The light of the night sky can be seen particularly on clear moonless nights, when the sky is never completely dark but appears as if a kind of twilight is lingering even though the Sun is well below the horizon. There is no doubt that some of the light of the night sky is due to the light of the stars, of the Milky Way and its bright nebulae. The contribution of all these is however, small and the greater part of the light is produced by our own atmosphere.

As happens in the case of aurorae, but in a more constant, calmer way, the Sun's rays, and particularly the ultraviolet ones, break up the molecules which exist in the upper atmosphere into atoms and these latter become ionized, or in other words lose electrons. The atoms excited by the solar radiation, emit visible light, which make the stars appear paler. It is clear that this light is therefore a handicap to the astronomer in his telescopic observations be it visual or photographic. In the case of photographic observations this light is particularly troublesome since it tends to fog the plates and therefore it limits the length of the exposure time. This is an additional reason why astronomers wish to work with telescopes above the terrestrial atmosphere. A wish which is increasingly becoming a reality.

The light of the night sky appears in various colours at all latitudes. Spectroscopic analysis shows that the spectrum is not a continuous one. Only short regions of the spectrum appear bright. A comparison with various luminous sources available in our laboratories tells us what gases and elements produce the light of the night sky in that layer of our atmosphere produced at a height of about 60 miles above the surface of the Earth, by the action of the Sun. Particularly noticeable is the green light which is due to atomic oxygen mixed with argon, under the action of electromagnetic discharges. The red light, also due to oxygen, is less intense. Sodium, which is so abundant in the terrestrial crust and in the oceans is found not only in the terrestrial atmosphere but in interplanetary space as well. It is true that its quantity is very minute but when we consider how extensive is space, we can understand how its atoms, excited by the solar radiation, can emit light in a small yellow region of the spectrum.

The spectroscopic analysis of the light of the night sky will show quite clearly the yellow radiation due to sodium, just as it can be easily observed in a terrestrial laboratory when we burn a pinch of ordinary kitchen salt (sodium chloride) in a Bunsen burner. Nitrogen, which is one of the main components of the terrestrial atmosphere, can be detected in the light of the night sky in combination with oxygen namely in the form of nitric oxide. These discoveries have led to some valuable experiments by means of rockets. A certain amount of sodium and nitric oxide has been launched into space to the height of that layer where the light of the night sky is formed. By using a few pounds of sodium, it has been possible to produce a very intense yellow glow which was visible at great distances. From the actual drifting of this luminous cloud it was found that at a height of about 50 miles above sea level, a wind was blowing with a speed of 186 miles per hour. At a height of 68 miles, the wind was blowing in the opposite direction with a speed of nearly 100 miles per hour. From these experiments it can be appreciated how useful the sodium clouds are in the exploration of the upper atmosphere. Similar experiments carried out in Italy from the missile base in Sardinia, have shown that nitric acid released by a rocket at a height of about 62 miles gave rise to a vivid burst of light which lasted 10 minutes and left a luminous trail over a wide part of the sky.

At heights which are a little less than those which emit the light of the night sky, at about 50 miles, we observe, sometimes, a peculiar type of cloud in our atmosphere. These noctilucent clouds, as they are called, are much higher than the common clouds, indeed they probably are the highest which can be formed, almost at the limit of our own atmosphere. At night these clouds can be seen in various colours. This is due to the fact that they are so high that they still receive the light of the Sun, even when the surface of the Earth is in darkness. The light they emit, which is bent by refraction, enables us to see them brightly throughout the night.

Phenomena similar to these as well as new phenomena of a much larger scale have been observed in recent years by astronauts orbiting around the Earth in artificial satellites at heights well above the layers where the light of the night sky and noctilucent clouds are formed.

4 · The Exploration of the Moon

The exciting race between the Russians and the Americans to be the first to set foot on our natural satellite is in full swing. The Russians were the first to score some success by crash-landing a space craft on the Moon on September 13th, 1959.

A few months earlier Lunik I, although it succeeded in escaping from the gravitational field of the Earth, missed the Moon and became a tiny planet, or rather a meteorite, revolving around the Sun. The same happened in the following month of March to the smaller American probe Juno. The next Russian space craft which was launched in September 1959, succeeded in crossing the neutral region of space where the terrestrial attraction equals the lunar attraction and then, gradually accelerating, ended its journey by crashing on the Moon just as meteorites fall on the Earth. The main difference, of course, is that meteors in their fall towards the Earth have to travel through the atmosphere. Its layers nearer the Earth are denser, and as a result of the friction encountered, meteors normally burn out partially or completely. In the case of the Moon, because of the absence of atmosphere, the space probe fell as a projectile and must have broken up on impact against the lunar rocks.

The unexplored regions of the Earth had to be discovered inch by inch with great difficulty, until it became possible to photograph these regions from aeroplanes. The topography of the Moon, on the other hand, is already well known in considerable detail, at least as far as its visible hemisphere is concerned. Because of a phenomenon known to astronomers as 'libration', it is possible to see also about one-fifth of the other hemisphere. What we can see and what we can photograph of the Moon with the help of the largest telescopes, reveals the nature of the ground. Its characteristics can be examined in great detail and we can conclude that on the Moon all the conditions which make life on the Earth possible are lacking.

The Moon, like the Earth, of which perhaps it was a part, is a daughter of the Sun. Because of its rather modest dimensions, the Moon has not been capable of maintaining around it either hydrogen or oxygen or any other of the gases which compose our own atmosphere. The birth of the Moon must have been cataclysmic, at least if we are to judge from the appearance of the lunar surface which is pitted by craters, wrinkled by mountain ranges and marked by crevices and by desert areas. These peculiarities are known to us in all their details since the early telescopic observations and their names do not necessarily correspond to their nature (Plate 15). Thus we have *Mare Nubium* where there is no sea nor clouds; *Oceanus Procellarum* where there are no storms; *Mare Pluvium* where no rain ever falls; *Mare Serenitatis* and *Mare Tranquilitatis* where desolation and death reign. The question often asked is what covers the surface of these extensive grey coloured areas, which show as dark patches at the time of full moon. It is believed that the surface of these areas is covered by volcanic ash. If we want to be exact we ought to specify that the configurations which have been called craters, are almost certainly craters. An observer who is well acquainted with volcanic landscapes on the Earth, such as the regions of Vesuvius, of Etna and of other similar volcanoes, must admit that there is a considerable similarity between these volcanic regions and those on the Moon.

Eruptions no longer take place on the Moon which, therefore, does not develop any energy in its interior as is still the case for the Earth. Towards the centre of the lunar disc near a range of mountains called *Carpathian* we can see one of the most typical craters known as *Copernicus*. This crater has a diameter of approximately 56 miles and has at the centre a mountain nearly

2½ miles high. Of great interest also are three craters which have been named *Ptolemaeus, Alphonsus* and *Arzachel*. These cover an area of several miles and they lie close together in such a way as to suggest to Galileo the name of 'cauda pavonis' as he compared these craters to the eyes which embellish the feathers of a peacock's tail.

The mountain ranges are not as extensive as those on the Earth. In addition to the *Carpathian Mountains* there are the ranges known as *Apennines* and *Alps* which outline the edges of the maria. The terrestrial observer who surveys the Moon through a telescope, will notice that in addition to the large craters of extinct volcanoes similar to those existing on the Earth, there are others of very small dimensions which are scattered at random and which are very reminiscent of the craters formed on the Earth by the impact of meteorites or by aerial bombardments during World War II. Today, therefore, there is a tendency among astronomers to believe that in the upheaval which took place at the time of the origin of the Moon, not only were there violent eruptions, which were to be a prelude to the inactivity to which the Moon was doomed, but probably also violent showers of meteorites.

Less than a month after the first Russian success another great event took place. On October 4th, 1959, the Russians launched Lunik III to which was allotted the task of taking photographs of the side of the Moon which is always invisible to us. This exploit was completely successful. It is true that the photographs of the unknown hemisphere of the Moon were rather poor when received on the Earth, nevertheless, in them it is possible to detect some of the configurations which had never been seen by man before. We may say that the question of the appearance of the invisible side of the Moon is by now almost solved. The photographs taken by Lunik III did not show great detail of mountain, craters and maria, because they were taken when the Moon was almost full. In such conditions the light of the Sun is almost overhead of the lunar surface and the shadows cast by the mountains and the walls of the craters are practically non-existent. The general surface appears of a whitish colour and the maria appear dark. We have already mentioned that on account of the small oscillations that the Moon undergoes in its motion around the Earth (librations) we can see between 10 to 15 per cent more of one hemisphere. This, together with the relative positions of the Sun and of the Moon and its supposed origin, suggested that the invisible face of the Moon could not be very different from the visible side well known to us. The invisible side must also be subject to sudden changes of temperature from more than 100°C. when the Sun is shining to − 100°C. during the lunar night. There must be a total absence of atmosphere, of water and of clouds in both hemispheres. We could not expect, therefore, any great surprise in the appearance of the other side of the Moon.

In the south-western region, which in the Lunik III photograph is not yet completely illuminated by the Sun, we can see configurations already well known to us; the elliptical dark patches of *Mare Crisium* and *Mare Foecunditatis*. Let us once again remember that no water exists in these maria. Progressing towards the centre of the disc we can detect new craters which in many ways are similar to those already known to us. In spite of the poor details of the photographs we can see that some of these craters have central peaks which probably were the cones of the eruption at the time when the craters were active. We have similar examples on the Earth in the craters of Vesuvius and Etna.

In the north-eastern region of the lunar disc there is a large surface uniformly illuminated which would therefore suggest that no maria existed there. Perhaps there may be mountains and craters but these would not be visible because of the inclination of the light of the Sun. There is, however, a dark patch elliptical in shape. Is this a 'sea' or a crater? Its dimensions are comparable to those of the crater *Clavius* which is on the visible side of the Moon, and from this we might suppose that this is a crater but of an irregular shape.

From the evidence of the photographs we must conclude that there are differences in the distribution of features such as maria and craters on the two sides of the Moon. This is not really surprising since, after all, the same thing is true about the Earth

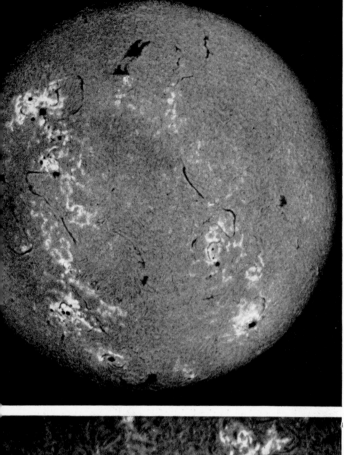

8.
Spectroheliogram
obtained in the
hydrogen line Hα
with a Lyot filter
at the Sacramento
Peak Observatory
(U.S.A.)

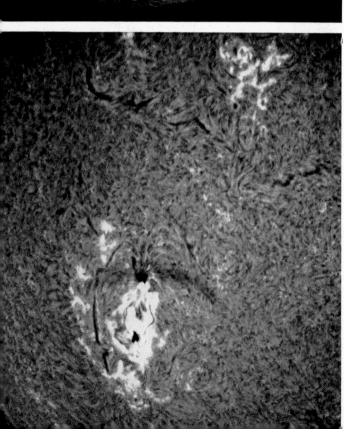

9.
Spectroheliogram
of flares,
filaments, spots,
taken with the
hydrogen line Hα
November 28,
1958, 13 h 55 min
U.T. at the
Anacapri station
of the Fraunhofer
Institute

10a. Photograph of a prominence in the light of Hα taken with a Lyot filter, June 12th, 1937 at 13 h 28 min U.T. (Meudon Observatory)

10b. Eruptive prominence taken in the light of Hα at the Sacramento Peak Observatory

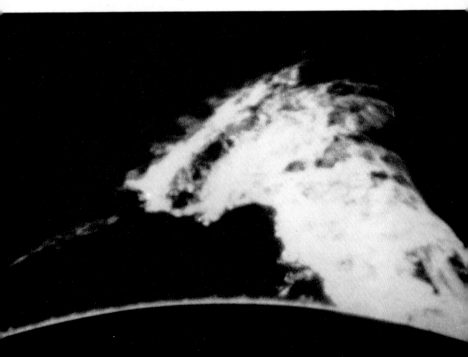

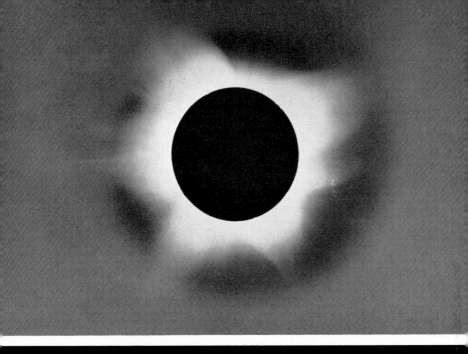

11. (*above*) The corona of June 19th, 1936. Polar type. Italian expedition to Sara, U.S.S.R.
12. (*below*) The corona of February 25th, 1952. Intermediate type. Italian expedition to Khartoum, Sudan

13a, b, c.
Various types of
aurorae

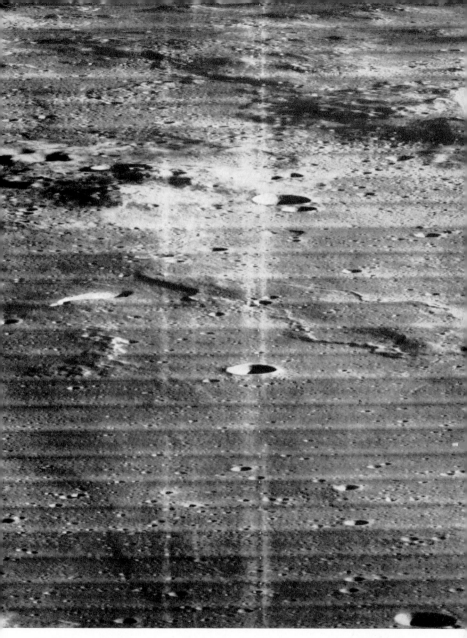

14. Photograph taken by Lunar Orbiter II. This is a view of the Marius hills area looking north 12 degrees east (north is at the top when the horizon is viewed at the top of the photograph). The large crater Marius (not shown) is east of the area shown

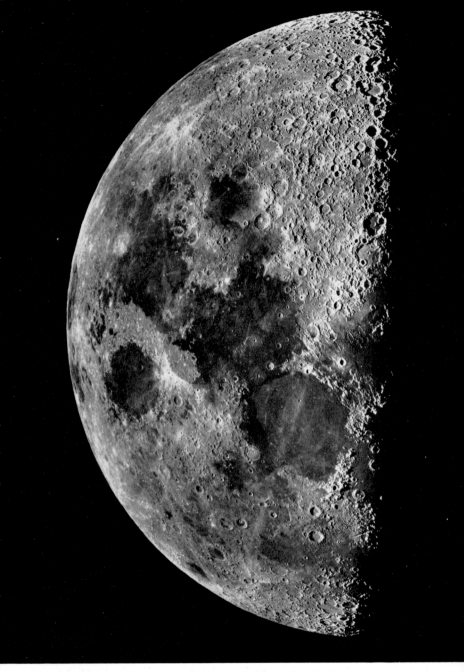

15. The Moon at first quarter (Moore and Chappell) Lick Observatory

16. (*Opposite*) The other side of the Moon—This wide angle view of the far side of t
Moon was received from Lunar Orbiter V, and shows surface features as small as 5
metres across. North is at the top and the terminator, or dividing line between sunli
and darkness, lies in the region of 105 degrees West longitude. The lighted porti
represents about one-fourth of the Moon's hidden side

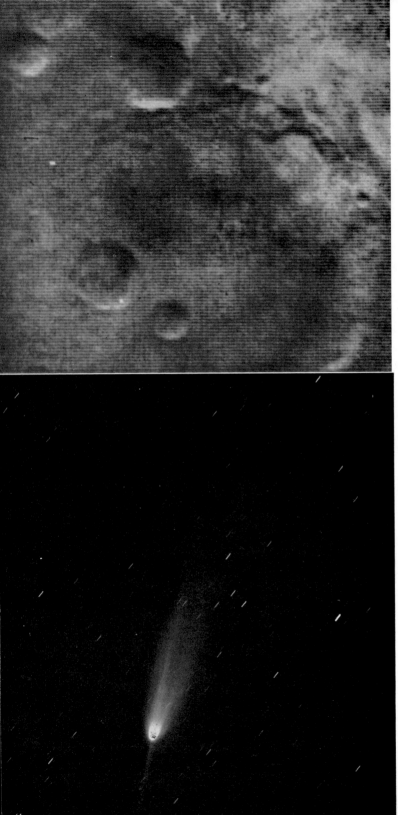

17.
Photograph of
Mars taken by
Mariner 4, with
green filter, at
at a distance of
7,800 miles from
the surface of the
planet, showing
Atlantis between
Mare Sirenum
and Mare
Cimmerium

18.
Comet Arend-
Roland (1956) at
Arcetri Observa-
tory

when we consider the distribution of land masses and of water in both hemispheres. Finally, in the lower right hand part of the photograph, there appears to exist a large sea to which has been given the attractive name of *Sea of Dreams*, a name which so far has not appeared in selenography.

On July 18th, 1965, another Russian spacecraft named Zond III was launched into space in an orbit which passed behind the Moon. Zond III took 25 photographs of the surface of the Moon which were of better quality than those obtained by Lunik III and confirmed the results already discussed above.

The photographs taken by Lunik III and Zond III have been followed by photographs of much better quality (Plates 14 and 16). Lunar Orbiter 3, launched in 1967 from the United States, obtained excellent photographs of the hidden face of the Moon from a height of about 900 miles from its surface. In May 1967 another American satellite, Lunar Orbiter 4, sent back to Earth photographs of the unknown face of the Moon which showed many new details (Frontispiece).

For years now astronomers have attempted to explain the features which exist on the surface of the Moon. Two hypotheses have been put forward which combined together may explain some of the phenomena observed. The first is the volcanic theory, according to which, at the time of its birth, the Moon underwent a violent period of volcanic activity as the result of which large craters were formed and rivers of lava ran over parts of the surface of the Moon. If we study the properties of the light of the Sun reflected by the various lunar regions, we can conclude that they are very similar to the general characteristics of the light reflected by volcanic ash and pumice-stone with a high content of silicone. Volcanic ash is grey, sometimes brown or bluish and similar hues are found in the faint colouring detected occasionally on the maria.

The second theory is the meteoric, which can be considered in some ways as a complementary theory. This theory originated from the difficulty of explaining purely by the volcanic theory some marked features of the craters, such as their shallowness and the fact that their edges are only slightly raised above the general surface of the Moon surrounding them. There is little doubt that with internal explosions of great violence and of great frequency the erupted material apart from forming rivers of lava, must have also been thrown up to great heights, almost like projectiles which on falling back on the surface of the Moon could have produced the craters. It is also possible, of course, that during the stage of condensation of the matter which formed the solar system, meteoric showers were formed which, falling on the Moon, would have produced the appearance and the shape of at least some of the craters.

In 1964 and in 1965 three American lunar probes, Ranger 7, 8 and 9, were crash-landed on the Moon, and as they were hurtling towards the surface they took close-up photographs of the visible side of the Moon. Altogether these three spacecraft have sent back 17,700 pictures of a very high quality, and as the point of impact in each case was a different one, a fairly large area of the Moon has now been photographed from a very close range. In the case of Ranger 8 and 9 the last photographs were taken from a height of a few thousand feet from the lunar surface, less than half a second before impact. They show a great number of craters and microcraters, some as small as $2\frac{1}{2}$ feet in diameter.

Scientists are already at work on the great wealth of information that these photographs provide, but it will be some time before we understand the full significance of all the data which have been obtained. It is hoped that these photographs will be able to tell us whether the surface of the Moon is covered by a deep layer of dust or not. Preliminary results obtained from these photographs have not yet given a definite answer to this question. It has been suggested that on the Moon there may be sedimentary processes which lead to particles becoming joined together by a kind of vacuum welding as well as by compression. The size and nature of the very small craters, which also appear to be relatively newer craters and which are shown in photographs taken by the Rangers, suggest a surface which does not present a great resistance to impact. This would appear to indicate that the surface of the Moon is not covered by a layer of dust, but rather consists of a kind of foam lava which is very brittle. According to the Russian scientists, the Moon is covered with a layer of porous but solid matter 20 to 33 feet thick. This substance, which the Russian scientists have called *lunite*, does not occur on the Earth and is probably composed of volcanic ash and slag, pumice and meteoric matter.

The next important step in the exploration of the Moon took place on February 3rd, 1966, when the Russians were able to soft-land on the Moon a capsule, Luna 9, which was able to send back to Earth photographs taken directly from the surface of the Moon. A few months later, on June 2nd, 1966, the Americans successfully soft-landed Surveyor I on the Moon's *Oceanus Procellarum*, at about 500 miles from the spot where Luna 9 had landed. Surveyor I is a capsule, or rather a small unmanned station, supported by a tripod. The capsule is about 10 feet high and weighs 620 pounds and carries a television camera which points to a mirror which can turn around in a horizontal plane and can also be tilted up and down. Three days later Surveyor I had already sent to Earth 2,503 photographs which were extremely clear and showed the immediate area surrounding the equipment as being covered with rubble, stones and rocks. Surveyor I survived for eight months on the surface of the Moon during which time it withstood eight cycles of extreme heat and cold. All together it sent to Earth a total of 11,000 photographs of the lunar landscape. The photographs were taken until darkness fell on the Moon and, from day to day, the changing altitude of the Sun above the lunar horizon gave varied length to the shadows cast by even small objects. Then night fell and Surveyor I was unable to send any further photographs. Scientists on the Earth, however, were waiting for the beginning of the new lunar day in order to obtain more photographs. By comparing photographs of the same areas it was hoped to detect whether any new craterlets had been added during this interval of time. When the Sun rose again, Surveyor I did not appear to respond for a few days to the instructions sent from the Earth until the Sun had charged the solar batteries. After that, photographs began to reach the Earth again to add to the 10,388 photographs already obtained during the previous period of daylight on the Moon.

The exploration of the Moon is now being intensified as the time is approaching when astronauts will be able to land on the surface of our natural satellite. Before this takes place, however, a great deal of knowledge is still required about the conditions prevailing on the Moon. Lunar Orbiter 3 had the task of reconnoitring areas which had been selected as likely places for a manned landing, such as the *Ocean of Tranquillity* which appears smooth enough to enable a landing to take place.

Perhaps the most exciting of the recent unmanned soft-landings carried out by the Americans was that of Surveyor 3 which provided evidence that a manned ship can land safely on the Moon and that astronauts can walk around without fear of sinking in thick dust.

Surveyor 3 was fitted with a robot claw at the end of an extendible arm, which could dig a small trench. Several diggings were carried out between April 21st and 23rd, 1967 and pictures of the operation were televised to the Earth. In digging a small trench, the robot claw met pebbles at a depth of 6 inches below the surface. Another experiment was carried out on April 27th, 1967, when a lump of soil was scooped and placed into one of the foot pads of the capsule and then photographed in black and white and by means of coloured filters. This gave us the closest look at the Moon's soil that we have had so far. The conclusions reached are that the Moon's soil has a consistency which is very similar to that of the soil of the Earth and that the bearing strength of the lunar surface is about the same as that of the Earth.

Slowly the detailed picture, of at least part of the Moon, is being built up. From this, much information will be obtained which will be of great interest to astronomers in the furtherance of their understanding of our natural satellite. The information will also be valuable for the choice of a suitable landing place for the great adventure when the first manned spacecraft will land on the surface of the Moon.

5 · The Exploration of Venus and the Wonderful Journey of Mariner II

Venus, the very bright planet, is always enveloped by a dense layer of clouds which prevent the questing human eye from investigating the surface of this planet. Among all the planets of the solar system only Venus and Mars, the nearest planets to the Earth, may conceivably have some form of life, although this life, if it exists, must be very different from that we know on the Earth. Venus has the advantage of having dimensions very similar to those of our own Earth, while Mars has a volume which is barely 2/10 of that of the Earth. This fact, coupled with the very rarefied atmosphere of Mars, implies that conditions are unsuitable for the presence of forms of life similar to those existing on the Earth. In the case of Venus, conditions are perhaps more favourable as far as its dimensions are concerned, but there are unfavourable conditions in other respects.

Venus, being nearer to the Sun than the Earth, performs its revolution around the Sun in about 225 days. The temperature on the surface of the planet, on account of it being nearer to the Sun, is higher than that on the surface of the Earth but it is not so high as to be a serious obstacle to life. There is, however, another factor which makes it difficult for life as we know it to exist on Venus. Seen through a telescope this planet shows phases similar to those of the Moon and the size is such that we should be able to detect any features on its surface. The very thick layer of clouds which constantly envelopes the planet, allows us to see only a very few and ill-defined markings which

149

appear to be of a transient character. Observing these markings for a few hours we conclude that the rotation of Venus is not a rapid one, comparable in fact with that of the Earth or of Mars, namely of the order of 24 hours.

The question of the rotation period of Venus, which is so important because of its bearing on climatic conditions, has not yet been answered. Schiaparelli, from the observation of the faint markings on the surface of the planet, deduced that the rotation of Venus was very slow, indeed that its rotation period was equal to its revolution period. More recent observations by means of spectroscopy, of radioastronomy and of photographs in the ultraviolet, seem to indicate a rotation period of about 14 days. If the rotation period was equal to the revolution period, then Venus would always present the same side to the Sun, just as the Moon does with reference to the Earth. This would mean that the hemisphere facing the Sun would be hot, while the other hemisphere would be extremely cold and this would certainly not be conducive to the development of any form of life on the planet. On the other hand, if the period of rotation was much shorter, then the situation would be different.

Observations made from the Earth with the help of a spectroscope, do not penetrate the thick layer of clouds surrounding Venus and enable astronomers to study only a very thin layer of its atmosphere. The spectroscope reveals the existence of carbon dioxide in this layer but shows no evidence of water vapour or oxygen which play such an important part in the development of organic life.

To the terrestrial observations can, nowadays, be added observations taken at considerable heights above our own atmosphere, by means of rockets and artificial satellites. Indeed it is even possible with space probes to make observations of Venus itself from a very close range. This was actually achieved by Mariner II. In this first attempt the space probe did not come very near to Venus but it is conceivable that in the future space probes may approach closer and even penetrate the thick layer of the clouds surrounding the planet, and thus supply us with definite information on the conditions existing on its surface.

Mariner II, which probably is still travelling around the Sun, was a great technological achievement, being a very complicated piece of equipment and perfect in every part. Two big wings carried the solar batteries which supplied the necessary current to the various instruments which, in turn, were controlled by radio signals sent from the control centre on the Earth.

The astronomers responsible for this Mariner project were faced with a very difficult task. They had to formulate the theory and carry out all the relevant calculations in order to determine the most suitable time for the launching of the space probe. In addition they had to calculate a suitable orbit for the spacecraft so that after a given number of days it could approach Venus without coming too near to it, otherwise it might fall on the planet on account of its gravitational pull. Right from the beginning, Mariner II would have to travel in an orbit somewhat similar to that of the Earth around the Sun, but crossing the orbit of Venus. Thanks to the precision and perfection of the methods of that branch of astronomy known as celestial mechanics, this became possible. In addition, the great progress reached nowadays in the instrumentation of spacecraft, made it possible to supply Mariner II with ancillary equipment which by means of instructions sent by radio from the Earth, could correct, if necessary, its course, vary its speed and alter its direction. In other words, we can today control spacecraft as easily as we can control ships and aeroplanes on the sea and in the air.

Mariner II was launched from Cape Kennedy in Florida on August 27th, 1962. Soon after its launching the spacecraft began transmitting valuable information on the conditions it was meeting in its journey through space and which was received by various stations on the Earth. Mariner II gave information relating to the solar wind (see page 57) and revealed the existence of weak magnetic fields in interplanetary space. The Sun, the Earth and probably all the celestial bodies which rotate around their own axes, are seats of magnetic fields of various intensity.

One of the problems which Mariner II had to investigate, was whether Venus had a magnetic field—assuming that it had a very slow period of rotation. For this reason a magnetometer was carried by Mariner II. This instrument showed that Venus does not appear to have a magnetic field of the intensity of that existing on the Earth. At the closest approach of Mariner II to Venus, which was 21,598 miles, the magnetometer would have been able to record a magnetic field if it existed. It is true that the magnetometer did not show the existence of a magnetic field, but we cannot exclude the possibility that the field may be so weak as to necessitate a much closer approach of the probe in order to reveal it.

During the 42 minutes when Mariner II was closest to Venus, several million data were transmitted which are being studied by electronic computers.

Among the most important problems to be solved in order to have some indication of whether life similar to that on the Earth might exist on Venus, are those concerning the existence of water vapour and oxygen and the determination of the temperature which exists in the various regions of the planet. Radiometers of various types were carried by Mariner II and they explored several points of the surface of the planet both in the part illuminated by the Sun and the part in the shadow. The results obtained show that the average temperature at the terminator, that is to say at the border between light and darkness, is about 425°C. and that no trace of water seems to exist. On the other hand, the temperature of the layers of clouds at a medium and at a high level is $-35°C$. and $-50°C$. respectively. When we combine these data with those obtained from our terrestrial observations, we conclude that Venus, although it has characteristics which are similar to those of the Earth, has a temperature which proves to be unsuitable for any form of life and that its sky is perpetually obscured by the thick layer of clouds in which carbon dioxide prevails.

Another very important result obtained by the spacecraft was the determination of the mass of Venus, namely of the quantity of matter which forms the planet. In the case of the planets of the solar system which like the Earth are accompanied by satellites, it is possible to determine the mass of the planets. This can be done by means of celestial mechanics by studying the perturbations which the satellites produce in the orbit of the planet around the Sun. Venus has no satellite, therefore the value of its mass which had so far been adopted, was based on the perturbations that Venus produced in the motion of Mercury. Mariner II acted as a satellite. The gravitational field of Venus produced perturbations in the orbit of Mariner II, by changing its course around the Sun when it passed in the neighbourhood of Venus. Since the position of Mariner II was continually checked from the Earth, it became possible to detect the variations in its motion and hence the perturbations produced by Venus. Thus the mass of Venus could be deduced with an accuracy much greater than hitherto. The mass of Venus is 8/10 that of the Earth.

In spite of the great success of the journey of Mariner II, we must admit that until such time as we can penetrate the thick shroud of clouds which surrounds Venus, we shall not be any nearer to solving the mystery concerning any form of life which may exist on it.

Translator's note:

On October 18th, 1967 a Russian probe, named Venus 4, made a soft landing on the surface of Venus. The preliminary results obtained from the data transmitted back from this probe indicate that Venus has an atmosphere which is 15 times as dense as that of the Earth. The surface temperature of the planet can reach very high values of the order of 530°F. and this would preclude the existence of any form of life.

The length of the period of rotation of Venus is a question which has not yet been finally solved.

6 · The Exploration of Mars — Canals, Deserts, Polar Caps and Possibility of Life on Mars

What will the first astronauts find when they first set foot on Mars? This is a question often asked in this age of space exploration. Mars is a planet in many respects similar to the Earth and capable, perhaps, of harbouring some very primitive form of life. It is strange, however, that some of the hypotheses which have already been discredited are occasionally still being discussed.

The Earth performs one revolution around the Sun in approximately 365 days, and Mars in approximately 687 days. The two planets find themselves nearest to each other when Mars is in opposition, that is to say when it is in a direction opposite to that of the Sun as seen from the Earth. Since the orbits of the two planets are not circular but elliptical, some oppositions are more favourable for observation than others. The minimum distance between the Earth and Mars in such circumstances can be nearly 35 million miles.

On account of this close approach and because the diameter of Mars is about half that of the Earth, it follows that even during these favourable oppositions the apparent diameter of Mars is only 25 seconds of arc. This is indeed a small angle and to realize how small it is we must remember that it is nearly 1/80 of the apparent diameter of the Sun or of the full moon. With the unaided eye the features of the lunar surface, such as the maria, the craters and the mountain ranges can only be seen as bright or dark patches. We can therefore understand how few details can be seen on a small disc which is only 1/80 of the lunar disc. When Mars is not at the most favourable opposition, its apparent diameter is even smaller and it is clear that then there is even less likelihood of detecting any details of its surface. In addition we must remember that the images are always affected and distorted by the ever present turbulence of our atmosphere and therefore it is understandable that the interpretation of the features shown by the surface of Mars is extremely difficult.

Many astronomers, both professional and amateur, have carried out observations of the surface of Mars by visual as well as by photographic means. They have recorded the patches of various hues and their variations during the seasons which follow each other like they do on the Earth, but in the case of Mars the duration of each is longer. During the favourable opposition which occurred in 1859, Father Secchi, director of the observatory of the Collegio Romano, was the first to discover, with the aid of a small telescope, some interesting features on Mars. He observed two red, narrow lines which were dark and permanent, between two regions which were lighter in colour. These were, possibly, deserts, and because of their appearance he named them *canali*. During the next favourable opposition, Schiaparelli, director of the Brera (Milan) observatory, continued the observations which are now so well known both for their number and their wealth of detail. Schiaparelli used with great caution terms such as canals, seas, continents and lakes, because they were reminiscent of similar terrestrial configurations, but in his work he always warned us of the fact that all these markings could well be interpreted in a different way.

There was already a precedent in the case of the Moon, where areas which had been called 'seas' were in reality deserts without any water and probably covered by volcanic ash. As the number of observers increased so the number of canals appeared also to increase. The canals seemed to form a wide network covering large regions of the planet. Caution was thrown to the winds, as often happens when a solution to a mystery is sought, and fantasy took over by attributing the presence of these canals

155

to work carried out by intelligent beings. Once launched on this course imagination knew no bounds, but the observations made at the end of the nineteenth century and the beginning of the twentieth century, finally disposed of all the hypotheses that these canals had been built for the purpose of irrigating the deserts.

The collapse of these hypotheses was caused by several factors. First by a detailed study of the physiological and psychological factors involved in the visual interpretation of details as minute as those shown by the surface of Mars. Secondly by using instruments of much greater aperture which enabled observers to use higher magnification. Thirdly by making use of photographic methods coupled with very sensitive emulsions, and finally by a comparison of the features of Mars with those of the Earth as seen on photographs taken at great heights above the surface of the Earth.

Telescopes used at the time of Schiaparelli were comparatively small. Through them the disc of Mars, even when magnified to the limit, appeared only slightly larger than the disc of the full moon as seen with the naked eye. As we have already remarked, such observations of the Moon by the naked eye or even by means of binoculars, show only patches of various brightness but certainly not details. In the case of Mars, observers using telescopes of small aperture, began to doubt the existence of canals and began to attribute them to optical illusions.

According to those observers who made almost geometric drawings of canals, these only appeared clearly in rare moments lasting at most for one-tenth of a second, which seems to confirm the theory that it was a case of optical illusion. This interpretation was strengthened by the observations made with large telescopes. Antoniadi, at the Meudon Observatory near Paris, used a 34-inch telescope while American astronomers had at their disposal the largest instruments in the world including the 200-inch at Mt. Palomar. All these observations seemed to confirm that the early observers had been misled by optical illusions.

Photography made a great contribution to the understanding of problems concerning Mars. At first the early photographs

156

were not very clear because of the smallness of the image and turbulence of the atmosphere, and because the grains of the photographic emulsions were far too large to allow enlargements to be made. Gradually, however, the production of fine grain emulsions, the use of filters of various colours and above all, the methods used by the astronomers of the Pic du Midi, gave very satisfactory results which confirmed modern visual observations. The method used by the astronomers of the Pic du Midi consists of taking many photographs of the planet in rapid succession and then of combining photographs which were not taken with too great an interval of time. This, of course, is essential because we must remember that Mars rotates around its axis like the Earth, in nearly 24 hours and therefore photographs taken at intervals of time greater than five or six minutes, could not be overlapped. Photographs, when overlapped and printed as one single photograph, show remarkable details, without the presence of the canals which the eye creates by integrating the various details.

While awaiting further exploration by spacecraft, we can put forward some plausible hypotheses concerning the other features of Mars by comparing the photographs of Mars with those of the terrestrial surface taken from great heights by the meteorological artificial satellites Tiros.

The Tiros satellites have been launched by the United States into orbits around the Earth in order to obtain a better knowledge of meteorological conditions prevailing on Earth. These satellites can give, among other information, pictures showing the distribution of clouds, the formation of cyclones and therefore also a picture of the Earth as seen from great heights when no clouds are present. The pictures, taken from a height of about 60 miles or more are of great value since they show us in what manner large expanses of water, deserts, mountain ranges, snow and ice and the various types of cloud, reflect the light of the Sun falling upon them at various angles. Since, after all, the light of the Sun is also that which illuminates the various features of the surface of Mars, it is permissible to make a comparison.

The Tiros satellites have taken photographs of the Mediterranean Sea, Egypt and the Nile valley, the Red Sea and the Sinai Peninsula, the whole of Italy from the Alps to Sicily and all the surrounding sea. The most striking fact which emerges from these black and white photographs is that the wide expanses of water appear very dark, while land appears with various degrees of light and darkness according to the vegetation which covers it, or whether it is a desert. Snow appears very white and bright and the clouds are also very bright, although, of course, the shape of them is different from what we are accustomed to see since the photographs are taken from above the clouds. The various degrees of brightness in the photographs are due to what astronomers call 'albedo' which is the ratio between the quantity of sunlight reflected by the surface of a planet and the quantity which actually falls on the planet. Thus, snow reflects the sunlight almost completely, nearly 8/10; clouds, on average, reflect 4/10; volcanic lava, which appears to be in such abundance on the Moon reflects about 1/10; while seas and lakes reflect only 5/100. This is the reason why the oceans appear so dark on photographs taken at great heights.

If we attempt a comparison between a map of the Earth and that of Mars drawn from the detailed and accurate drawings of many observers, or from photographs, we can perhaps see a faint similarity between the two maps. The southern hemisphere of Mars, up to medium latitudes, seems to be covered by regions of varying degrees of darkness. As we approach the equator of Mars and we reach those latitudes which on our own Earth we call tropical regions, we find an increase of dark regions which near the equator end in elongated shapes, remotely similar to our continents, namely South America, Africa and Asia. When we approach the northern latitudes, the surface of Mars becomes increasingly lighter in colour, and it is in these regions that have been drawn the greater part of the network of canals which actually can be resolved into a large number of small dots.

The whole of the martian landscape is dominated by the polar caps at the poles of the planet. With an astronomical telescope which inverts the images, we can see the south pole shining very brilliantly on the upper part of the disc. This polar cap appears quite extensive and slightly higher than the general level of the surface of the planet when the southern hemisphere is in its winter season. The northern polar cap is less conspicuous because it is visible only when the planet is at opposition near aphelion, namely when it is further from the Earth. Nevertheless, the presence of the polar caps is beyond doubt and this encourages us to continue the comparison between the appearance of Mars and that of the Earth. We can therefore attempt to put forward some reasonable hypothesis concerning the interpretation of the features observed on the planet.

The less dark regions of Mars, more common at high southern latitudes, when studied with a low magnification appear to be of a pink or reddish colour and when we use high magnification their colour appears to be nearer to orange. Probably these are lands almost completely lacking in vegetation which are very arid continental regions consisting of deserts. From the high latitudes as we proceed towards the medium latitudes, we find the very dark regions which have been named 'seas' by analogy with the terrestrial seas as they appear when seen from a great height. Spectroscopic observations show that the atmosphere of Mars is extremely rarefied and that there is scarcely any water vapour. This means that both clouds and precipitation must be rather uncommon so that we must conclude that no seas exist on the planet. The regions known as seas must, therefore, be analogous to the terrestrial regions which are covered by vegetation. This hypothesis is supported by the change in colour of these areas. No doubt the shape of these dark patches does not change but is subject to seasonal variations and must hence have a variable albedo.

The region that more than any other attracts the attention of observers is that known as the 'eye of Mars' which consists of a yellowish-orange area, slightly oval in shape and which was named *Thaumasia Felix* by Schiaparelli. In the middle of this region there is a dark patch which is almost circular and which is known as *Solis Lacus*. Here again there is no question of it being really a lake but rather an area in which some sort of vegetation exists which changes its colour with the changing seasons of Mars.

If we wish to push further the possible analogy between terrestrial and martian features, we could imagine a photograph taken from a great height of the region of Indonesia with all its numerous islands grouped in a semicircle around Borneo, which in our comparison would represent *Solis Lacus*. There is little doubt though that the vegetation of Borneo is much more lush than the very thin vegetation of *Solis Lacus*. Because of lack of water the *Thaumasia Felix*, or 'Land of Wonders', is probably no more than a sort of desert.

The polar caps, which are so similar in appearance to the terrestrial regions around the poles, show the only known form of precipitation on the planet. We must remember that the polar caps on Mars are only covered by a thin layer of frost, which under the action of the summer sun quickly disappears leaving the ground uncovered and subject to seasonal changes in colour. Rare clouds at times cover some of the regions of the planet but thicker veils of clouds may suddenly cover very large regions of the planet as happened at the last favourable opposition in 1956. These veils can be interpreted as being clouds of sand raised by strong winds.

Before the time was ripe for the sending of a spacecraft into the neighbourhood of Mars, a very important experiment was carried out at Princeton University. A large telescope was carried aloft by a balloon and this telescope was capable of obtaining photographs of Mars from a height of nearly 15 miles, free from the turbulence of the lower layers of our atmosphere. At the time of these experiments Mars had an apparent diameter of about 8 seconds of arc. In 1969 when the Earth will be nearer to Mars, the apparent diameter of the latter will be considerably greater, nearly 25 seconds of arc. This will, of course, be a suitable time for the observation of the details of its surface. Nevertheless, experiments with telescopes carried by balloons and with spacecraft can be carried out in preparation for the favourable opposition in 1969.

The balloon which lifted the telescope, Stratoscope II, consisted in effect of two balloons, a smaller one mounted above the main and larger balloon. When the balloon is launched the larger balloon is folded and has the shape of a cylinder. At a height of 3,600 feet, the helium contained in the smaller balloon expands and inflates the main balloon because of the reduction of air pressure at that height. The system is very stable and can be controlled even in winds with a speed of 25 miles per hour. This means that in order to launch the balloon there is no need to wait for a calm day.

The telescope, weighing approximately three tons, is a real masterpiece of modern technology. Its complex structure is nearly 23 feet long and its main optical part consists of an astronomical mirror 36 inches in diameter and weighing approximately 440 pounds.

The size and weight of this telescope are so great that it is astonishing that a balloon can lift such an instrument. That it is possible to control the work of the telescope and to dictate its settings from the Earth may appear even more surprising. Mars, although a bright object, is after all only a point in the sky and the telescope has to find it, and once it is found, the instrument must be able to hold it in its own field, neutralizing both the motion of the planet and that of the telescope itself.

Stratoscope II, in the night following the launching from Texas, was instructed to observe Mars, the Moon, Sirius and Betelgeuse from a height of about 15 miles above the Earth. Next morning the telescope landed safely by parachute after a flight of nearly six hundred miles, near Pulaski in Tennessee.

The observations of Mars carried out by Stratoscope II seem to indicate that in the martian atmosphere there exists a very small quantity of water vapour.

This is very important when we are considering the possibility of the existence of any life on Mars similar to ours. In addition, it confirms the observations made from our terrestrial observatories. If we consider that the atmosphere of Mars may have undergone processes similar to those which took place on the Earth, it must originally have consisted of water, hydrogen and ammonia. Water was converted partly into oxygen and hydrogen as a result of photochemical reactions in the upper atmosphere. Once the hydrogen had disappeared because of the weak gravitational force on Mars, ammonia was converted into nitrogen and methane gas into carbon dioxide. In the course of these changes organic compounds are produced and life, as we know it, develops. Because the lighter gases escape faster from the surface of Mars, development on this planet must have taken place at a much faster rate than on the Earth.

7 · Appointment with Mars

On November 28th, 1964, the Americans launched a spacecraft, Mariner IV, with the object of passing close to Mars and sending to the Earth photographs of the planet.

As can be seen in figure 8 both Mars and the Earth describe orbits around the Sun and on the date of the launching they were on the positions marked.

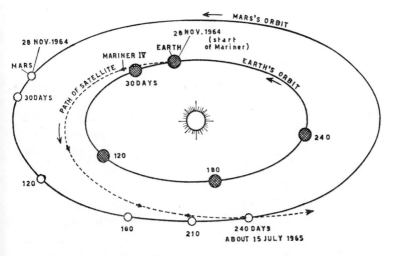

Fig. 8. Orbits of the Earth and Mars and path of Mariner IV

The Earth moves around the Sun at a velocity which is higher than that of Mars. In fact the Earth completes one revolution

163

around the Sun in approximately 365 days, moving with a velocity of 18 miles per second, while Mars takes approximately 687 days to complete its revolution around the Sun moving with a velocity of 15 miles per second.

Mariner IV was launched by a two-stage rocket, first into a parking orbit which was at a height of 115 miles and then, by firing the third stage, it was brought to its final orbit which was beyond that of the Earth and which was to cross the orbit of Mars at a determined point. The calculations required for determining this point are based on the laws of celestial mechanics, and nowadays are considerably simplified by using electronic computers. Astronomers, in considering the relative positions of the Earth and Mars in their orbits, decided that the most suitable time for such an experiment was towards the middle of November. After this date the Earth and Mars would not be again in a similar favourable position for 780 days, so that the experiment could not be repeated until January 1967.

Once the spacecraft was launched on its long journey, a continuous check had to be kept on its position, because perturbations could affect its predetermined orbit, and in such cases a correction to the course could be applied by means of the firing of small rockets from Mariner IV and controlled from the Earth.

The aerial carried by the spacecraft which was to be used to transmit to the Earth the information collected, had to be constantly directed towards the Earth. This was done by instructing the spacecraft to set its course by a star, in this case Canopus. Once the photoelectric equipment carried by Mariner IV had found this star, the aerial would automatically be directed towards the Earth. In addition, the photoelectric devices had also the task of maintaining the course of the spacecraft. At last, on July 15th, 1965, Mariner IV reached a suitable distance from Mars as predicted and began transmitting photographs.

Naturally these photographs were anxiously awaited on the Earth. For the first time it was going to be possible to have some real information about the features of the surface of Mars. The first photographs were rather disappointing as they simply showed slight differences in colour of some of the martian regions. Later photographs, however, revealed unknown and un-expected features of the red planet.

The limited region of Mars photographed by Mariner IV revealed the existence of a large number of craters very similar to those which exist on the Moon and on the Earth (Plate 17).

There is little doubt that if in future we could obtain photographs of the surface of Mars taken at an even closer range, we would be able to compare the distribution and the number of craters on these three celestial bodies which are similar from the point of view of their physical constitution. At the time of their formation and later, during their evolution, these three bodies, namely the Earth, the Moon and Mars, underwent changes which have given different final results. For the time being we can say that if photographs of volcanic regions of these three bodies were all taken from the same distance, they would show very little morphological difference, even if the actual number and dimensions of the craters are different. As a confirmation of this, it would suffice to compare photographs taken by aircraft of regions like those of Vesuvius or of the volcanic lakes of central Italy, with the photographs obtained by the Rangers and those obtained by Mariner IV.

The discovery of the existence of craters on Mars so similar to those of the Moon and of the Earth, came as a great surprise to astronomers who believed that their visual and photographic observations had supplied them with an adequate knowledge of the features of Mars. We must not think, however, that the general characteristics carefully studied, drawn and photographed by the astronomers in the past are to be considered completely wrong. After all, this has certainly not happened in the case of the Moon. Although it is true that astronomers had no idea of the details revealed by the photographs taken by Mariner IV, we ought to remember that hypotheses had been put forward on the existence of volcanic activity past or present on Mars. Some years ago Lyot studied the polarization of the light of Mars, that is to say the special characteristics of the light of the Sun reflected by Mars, and he found that the polarization was similar to that observed in the case of the Moon. He concluded that the surface of Mars could well be covered by a dust similar to that covering the Moon, namely volcanic ash. Only about 10 years ago, McLaughlin also suggested that it was premature to exclude the existence of volcanic action on Mars, and he reached this conclusion by carrying out special observations of the surface of Mars.

The hypothesis of the presence of vegetation does not exclude volcanic action since we have evidence of both occurring on the Earth. The configuration of the dark regions of Mars, improperly called maria, which are funnel-shaped from the southern to the northern hemisphere, suggested to McLaughlin the existence on the planet of a system of winds having a well-defined circulation, in some ways similar to that existing on the Earth. This may explain the appearance of phenomena very like sand storms, which were so conspicuous during the 1956 opposition.

The sharp ends of these dark regions were, according to McLaughlin, volcanoes whose ash was carried by the winds towards the southern hemisphere. The changes observed in 1926 in *Solis Lacus* could be explained by larger volcanic eruptions than the terrestrial Krakatoa eruption, because very extensive areas of the disc of the planet were covered by dark clouds for several weeks.

Unlike the Moon, on which no volcanic activity is present, Mars may still have some activity, and in any case we must remember some important phenomena concerning Mars. Like the Earth, Mars has its seasons, although they are twice as long as those on the Earth because of the greater length of the martian year. The existence of seasons has been established by the formation of snow or frost at the polar caps which gradually melts away with the approach of summer. With this thaw, there is a general tendency for an increase in the dark regions and a change in colour from the poles towards the equator. Therefore some form of vegetation may well exist on Mars.

One thing is certain, Mars is not a dead world like the Moon. How much life there is on the planet and how it has developed are questions which will probably soon be answered.

166

8 · Jupiter Observed by Television and by Radio

On October 2nd, 1933, the light of the Moon was received through the original telescope of Galileo at Arcetri and was used to switch on the lights of the world exhibition held in Chicago and opened by Marconi.

This in itself is not an extraordinary feat. Photoelectric devices are used to transform light into electricity and this can reach a destination anywhere in the world without need of wires. On arrival the electric current can be used to close an electric circuit, which, in the case described above, was the illumination circuit of the exhibition.

In 1954, more than 20 years later, the image of the planet Jupiter obtained with one of the telescopes of Greenwich Observatory, was transmitted by television to all the receivers which were within reach of the transmitting station of the B.B.C. Here again there is nothing mysterious about this. We are all accustomed nowadays to see images transmitted by radio, and in the particular case in question the image transmitted was that of a celestial object obtained by means of a good telescope. The only difficulty in transmitting images of celestial objects is to find a suitable night when climatic conditions and the seeing are good enough to avoid causing too much distortion of the image.

Jupiter is the largest planet of the solar system. Its diameter is nearly 11 times that of the Earth so that nearly 1,300 earths would be required to make a sphere as large as Jupiter. If, however, we compare it with the Sun, or with those stars which are even larger than the Sun, then even Jupiter becomes an object of negligible size. It is the fate of humans to find themselves surrounded by objects which appear infinitely large or infinitely small so that in this case, when we call Jupiter a giant planet, it is only in relation to the other members of the solar system.

Jupiter seen through a telescope under moderate magnification is really fascinating. The disc of the planet appears crossed by bands of various hues, parallel to its equator. In these bands we can detect interesting details. One of the most remarkable of these is a large spot, red in colour and oval in shape, which can be seen in the southern hemisphere of the disc. In addition, Jupiter appears noticeably flattened at the poles. This is a consequence of its rapid rotation since the planet completes one rotation around its axis in a matter of only 10 hours.

Often, near the disc and in the plane of its equator, we can see the four greater satellites which were discovered in 1610 by Galileo who named them the *Medicean Stars* in honour of his patron the Medici.

Like Mars, Jupiter is for us a mysterious object and it is natural that we should ask ourselves what are its physical conditions and whether it could harbour life similar to ours. The possibility, of life as we know it, can be excluded for the reason that Jupiter has a density equal to that of the Sun, that is to say just a little more than the density of water, while the density of the Earth is nearly five times greater. From this it is clear that Jupiter is without a solid crust. Another fact which supports the statement, is that like the Sun, the various regions of the planet have a different velocity of rotation. Some regions rotate faster than others.

The 'Red Spot' seems to be an obstacle to the circulation of vapours or clouds which form the coloured bands or belts. When these arrive near the Red Spot, they seem to be unable to cross it and they are compelled to go around it. The whole of the Red Spot, which is 30,000 miles long, appears to move as a ship at the mercy of the waves, perhaps because of the relative motion of the vapours which surround it. Generally, as its name implies, the Red Spot is bright red, but at times it becomes grey or

168

whitish and can hardly be detected against the background of the surface of the planet.

From what we have said so far, it must be obvious that Jupiter must consist of a liquid or gaseous substance and be surrounded by a thick atmosphere, consisting of definite layers of clouds which are compelled to remain in the same regions on account of the rapid rotation of the planet, just as our clouds are affected by winds. It is possible that what we see on the planet is the uppermost part of great eruptions emanating from its interior. The spots of various degrees of darkness, gradually take the appearance of long trails of smoke not unlike the pine-tree shaped clouds which are produced on the Earth by volcanic eruptions and which expand and then are dispersed by the upper air currents. The Red Spot could either be a solid body, some think solid hydrogen, floating in an ocean of gas or else the bottom of a volcanic region. So far nothing definite is known about it.

In addition it was discovered a few years ago that the higher part of the atmosphere of the planet consisted of ammonia and methane. All this precludes the existence of human life and even any other form of life known to us. On the other hand it could be suggested that on Jupiter there exist elements from which some form of life might develop in the distant future.

In July 1955 a group of American scientists announced that they had detected radio waves emanating from Jupiter, just as in early years radio waves had been detected from the Sun and other parts of the sky. This announcement required confirmation as often this type of observation can be masked by radio waves of terrestrial or atmospheric origin. Confirmation of this new phenomenon came from the observers of the Carnegie Institute in Washington, who had a very powerful radiotelescope in Maryland. The aerial system of this radiotelescope consists of an array of dipoles arranged in two lines each of which is over half a mile long and at right angles to the other. Such dimensions are necessary in order to be able to locate in the sky sources of very small angular dimensions like the stars, which to us appear as points of light. The aerial system is directed to a given region of the sky and it can receive signals from the celestial bodies which, on account of the rotation of the Earth, appear to pass by in that part of the sky to which the aerial is directed.

The Carnegie observers detected a source of radio waves in Gemini, which consisted of strong noise as if produced by irregular bursts of signals, in the same way as in our own radio receivers we can hear a strong crackling noise when a thunderstorm is approaching. When the observers located the direction of the source and the time of arrival of the signals, they noticed that these signals, although irregular, lasted for about six minutes. This interval of time was exactly the time taken by Jupiter to cross the aerial beam, as a consequence of the rotation of the Earth. The wavelength used for these observations was about 13·5 metres. In the case of the Sun, radio waves are received on a wavelength of 10 metres and these are emitted by the corona in which atoms are in a high state of excitation.

It is not really surprising that stars and the Sun itself should emit radio waves in the centimetre and metre wave-band because of their extremely high temperatures and as a result of electromagnetic phenomena. From the intensity of the emission of radio waves from a star it is possible to determine the temperature, which is not to be understood in the common sense of the word, but rather as the degree of excitation of the gases of which the stars are composed. Let us use a terrestrial analogy in order to understand this question. In our clouds we certainly do not have a high temperature and we have proof of this by the presence of ice crystals and the formation of hail. When electromagnetic phenomena occur in these clouds in the form of violent electric discharges, we can then talk of a high temperature of excitation followed by emission of radio waves which we call atmospherics, and which interfere with our ordinary radio programmes.

Jupiter is a much larger planet than the Earth and is also further away from the Sun. The quantity of heat it receives from the Sun is therefore much smaller and can be easily calculated. The temperature of the planet is extremely low, so low that no life, as we know it on Earth, could survive. Since, however, observations indicate that Jupiter is liquid and gaseous, it could well contain internal energies much greater than those which undoubtedly exist in the interior of the Earth. We actually have a proof of this by observing the peculiarities of Jupiter's surface which seems so disturbed. In addition this planet has a rapidly rotating atmosphere whose velocity, in the equatorial regions, reaches about 466 miles per minute. The unexplained complex phenomena which are observed on the surface of Jupiter could be interpreted as manifestations of violent periodic winds which carry matter ejected from the interior of the planet.

On the other hand, by means of the delicate instruments available nowadays, we can measure the surface temperature of Jupiter and we find that it is slightly higher than that obtained theoretically by calculations based on the distance of the planet from the Sun. It is very difficult, therefore, to explain how on Jupiter there could exist phenomena of such power capable of emitting radio waves which can be detected by our receivers. We have already mentioned that the word 'temperature' can have various interpretations. It is possible to assume that although the average surface temperature of Jupiter, both calculated and measured, is extremely low, nevertheless there may exist local phenomena and strong disturbances which may reach the surface from its interior. Disturbances may be formed in its atmosphere and may give rise to electromagnetic phenomena similar, in some ways, to those occurring in our own atmosphere, but on a much greater scale. This is not improbable if we consider the physical conditions of the planet.

Jupiter is not the only planet which emits radio waves. With the more powerful radiotelescopes which have been built in recent years, radio waves have been detected also from Venus and from Saturn.

9 · Saturn and its Rings

It is strange that among the great abundance and variety of objects which we can see in the sky, only Saturn should be surrounded by its wonderful system of rings. Perhaps we ought to remember that even if in other solar systems there existed planets like Saturn we would not be able to see them because of their relatively small dimensions and their enormous distances from us.

Jupiter and Saturn are the two giant planets of our solar system. Saturn's diameter is nearly 10 times that of the Earth. In spite of this and because of its distance from us of more than 600 million miles, the disc of Saturn appears to us only as one-hundredth of the disc of the full moon, even when Saturn, in its revolution around the Sun, comes nearest to the Earth. It is clear that even in the most favourable conditions it is necessary to have the help of a telescope with a reasonable magnification, in order to study the rings, the possible features of the surface of the planet and their variations. Any observer who has attempted this, realizes soon enough that the magnification cannot be pushed beyond a certain limit dictated by the turbulence of our own atmosphere. Only when the turbulence is at a minimum can we hope to detect those fine and rather vague details shown by the disc of Saturn, both in the course of the seasons and in the course of the various positions that the ring can assume.

The first important fact that astronomers find is that Saturn does not rotate like a solid body; therefore the planet must be

172

liquid or gaseous. Saturn completes a rotation around its axis in only 10 hours and shows a considerable bulging at its equator. With the aid of the spectroscope it has been possible to determine that the gases at the surface of the planet are methane and ammonia at a temperature of about −150°C. and that therefore the ammonia must be in a solid state. Observations show formations of clouds which are variable and which seem to be carried by strong currents, perhaps caused by a residual of internal heat which rises to the surface of the planet.

From the observations that we can make of the surfaces of both Jupiter and Saturn, there is very little doubt that very large phenomena of atmospheric circulation must exist in the atmosphere of both planets. These phenomena may well be accompanied by electromagnetic manifestations, as happens in the case of our own Earth, and this seems to be confirmed by the recent detection of the emission of radio waves by both planets. Even in the field of planetary observations radioastronomy offers new methods of investigation which may in due course lead us to a better understanding of the phenomena which occur on the planets of the solar system.

Ruggieri, an amateur astronomer from Venice, was able to carry out some observations of Saturn during its opposition in May 1955. He made his observations at the Merate Observatory with the same Merz-Repsold instrument of 19 inches aperture, used by Schiaparelli for his observations of Mars. At the time of the opposition, Saturn was turning its northern hemisphere towards the Earth. The ring was seen fully open, so that the northern face of the ring was clearly visible. Ruggieri could observe, in the course of several suitable nights, the details of the equatorial zone with the exception of the part next to the south equatorial belt, which was hidden by the innermost ring.

Without going into all the details of the careful observations made by Ruggieri, we can say that on the whole, at the time of the 1955 opposition, Saturn showed very fine and complex details as well as a total absence of well defined spots which could have helped to determine its period of rotation. The absence of spots and of variable details may well be an indication of relative calm in the atmosphere of the planet. Perhaps, if we want to push our hypotheses a little further, we may suggest that this was the result of the quietness of the Sun at the time, which, only at the beginning of that year was awakening to its 11-years' cycle of activity.

In the observations of the rings, Ruggieri was able to call attention to a rather rare phenomenon of the external ring. There seemed to be some very pale radial lines on the 'ansae' and they appeared to originate from Cassini's division and to converge towards the centre of Saturn. Similar phenomena had already been seen by earlier observers but it is not easy to put forward a satisfactory explanation. Perhaps it is an effect of illumination of the rings by the diffused light of Saturn itself. During the observations made by Ruggieri, it was once again noticed, as it had already been noticed before, that the transparency of the innermost ring decreased towards the interior. The variation in transparency of this ring can be seen by observing the disc of Saturn which is visible through the ring. On the night of June 25th, 1955, the ring showed areas of a varying degree of luminosity and this was taken to be an indication of its various condensations.

If we wish to explain the nature and the constitution of the rings, we must take into account the results of observations made with powerful instruments. The system of rings which surrounds Saturn, has since Galileo's days been a subject of many observations and theoretical research. From all these it became apparent that the ring was not a single ring, but is composed of a number of concentric rings, the innermost of which is nebulous, almost like a veil situated between the planet and the most brilliant ring. Of the two main rings, the more external has a diameter of 170,000 miles, and the most important division which follows and which was discovered by Cassini, is nearly 3,000 miles wide.

Theoretical considerations showed that the system of rings could not consist of a continuous solid or liquid surface, because at the slightest disturbance a solid ring would break into pieces which would fall on the planet, and a liquid ring would have divided into many parts. Thus, even before any observational evidence became available, it was suggested that the stability of the ring was due only to the fact that it was composed of a swarm of small satellites or meteorites. This hypothesis was completely confirmed by observations of the intensity of the light of the ring and by spectroscopic observations. The observations of the light confirmed that the light which reaches us from Saturn's rings is purely sunlight reflected, as a whole, by the particles which form the ring. The spectroscopic observations showed that the external parts of the ring revolve around Saturn with a velocity which is smaller than that of the inner parts, just as if it was a question of several satellites rather than of a continuous and solid ring.

Phenomena of great interest and which throw some light on the physical properties of the system of rings, are observed about every 15 years, when the rings are seen edgeways and to the terrestrial observer, almost seem to disappear.

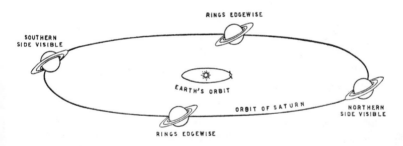

RINGS EDGEWISE

SOUTHERN SIDE VISIBLE

EARTH'S ORBIT

ORBIT OF SATURN

NORTHERN SIDE VISIBLE

RINGS EDGEWISE

Fig. 9. Orbit of Saturn and appearance of rings as seen from the Earth

Saturn completes one revolution around the Sun in $29\frac{1}{2}$ years. Every 15 years the plane of its rings passes through the Earth and since the rings are very thin, being only about 12 miles thick, they seem to disappear from sight almost completely for a few days. The plane of the rings takes almost a year to cross the orbit of the Earth and the latter, in this interval of time, may cross the plane of the rings once, or three times, according to the relative positions of Saturn and of the Earth in their relative orbits (fig. 9).

Observations made in 1937 and in 1950, when the ring was last seen edgeways, confirmed the fact that in reality the ring does not disappear completely but that a thin dark line can be seen crossing the face of the planet. Beyond this, irregular condensations can be detected. During 1966 the Earth passed three times through the plane of the rings. This was the first triple passage in almost half a century.

The system of rings could be observed closing gradually until it appeared edgeways only to open up again. In the second half of February the planet was visible in Aquarius soon after sunset. The northern part of the ring could be seen, then the Sun approached the position of the planet and on March 10th, Saturn, being in conjunction, was invisible. On April 2nd, the Earth crossed the plane of the rings for the first time, but because of the position of Saturn in the sky, the disappearance of the ring could not be observed. Towards the end of April, Saturn became a morning star in Pisces, and the ring began to open slowly so that from mid-June to mid-July the southern face of the ring system could be seen. After this the ring, once more, began to close until the end of October. On October 29th the Earth crossed the plane of the rings for the second time. Finally, on December 17th, for the third time in the year, the Earth crossed the plane of the rings, and the third disappearance of the ring took place. After this the ring began to open again and for the next 15 years, observers from the Earth will be able to see the southern face of the system of rings.

Many questions concerning the constitution of the rings remain still unsolved, such as that of the size of the fragments which compose them. Probably they consist of meteorites, since a very fine dust would not be able to withstand the pressure of the solar radiation. Recent investigations carried out in the United States concerned the analysis of the light reflected by the various rings. The results of these investigations indicate that the particles of which the rings are composed, are either encased in ice or are simply ice crystals, the existence of which can be easily explained by the extremely low temperature of the planet which is so far from the Sun. The inner ring, called the 'crape' ring, is almost certainly composed of matter of a very low density. Perhaps it is here a question of a very fine dust intermingled with gas, similar to the composition of the zodiacal light or of the tail of comets.

Now that astronomers possess more powerful instruments and more refined means of research, they will probably succeed, at the time of a favourable opposition of Saturn, in determining the nature of the dust forming the ring and in explaining the actual presence of methane and ammonia which are so abundant on Saturn itself.

10 · The Possibility of Life on Other Planets

The question is often asked whether any of the other planets of our solar system are inhabited by beings. If such beings existed and were not human, to what extent would they have developed towards similarity with man and how would their life span compare with ours?

We know, or at least we think we know, the past history of man on the Earth, and as far as the future is concerned, which in part may well depend upon humanity itself, it is ruled by a divine will and we cannot predict it. It is possible that if life, as we know it, did exist on some of the other planets of the solar system it might be either at the stage at which life on Earth was thousands of years ago, or it might be at a much more advanced stage of evolution, such as man may reach on Earth in thousands of years to come.

It may also be possible that beyond our own solar system, the infinite number of suns which can be seen in the sky, may have a family of planets similar to those which belong to the solar system. In this case it is obvious that the possibility of the existence of forms of life is much greater. While we can determine the presence of such planets belonging to other suns and calculate their dimensions, we cannot see or study them, because of their enormous distances from us.

We can, on the other hand, discuss the conditions prevailing on the planets of our own solar system with reference to the possibility of life on them. It is strange that we are forced to reach the conclusion that the Earth is the only planet of the solar system suitable for the maintenance and support of human life. This conclusion is reached from the study of the physical conditions of the planets and from the visual observations of their surface carried out with our large and powerful instruments.

Many theories have been formulated in the past and new ones are still being put forward concerning the origin of the solar system. Among these theories there is one in particular which is accepted in its general outline. We are referring to the theory put forward a long time ago, by Kant and Laplace, according to which the Sun was surrounded by a very thin gaseous envelope, lens shaped, which constituted the solar nebula. This nebulosity condensed slowly, in the remote past, perhaps 2,000 or 3,000 million years ago, into matter from which the planets were formed as we now know them. From this theory it is obvious that all the planets, like all celestial bodies, are composed of the same chemical elements which exist on the Earth. In addition, the theory of Kant and Laplace brought up to date, explains also the distances between the Sun and the various planets and how the elements, of which the planets are composed, can enter into chemical combinations which are different according to the dimensions of the planet.

Planets such as Mercury, Venus and Pluto have never been heavy enough to be able to retain much of the gas of the original nebula and as a result have remained with a rocky surface and a very thin atmosphere. Jupiter and Saturn, which are much bigger, were able to retain a considerable gaseous envelope before this could disperse into the surrounding space.

Starting an hypothetical journey from the Sun, a journey which will never be possible for man both because of the great distance involved and because of the heat of the solar surface, we would first meet Mercury. On account of its proximity to the Sun, this planet can only be seen at twilight or during the hours of the day and hence it is very difficult to observe. Mercury has a diameter, mass and volume which are considerably smaller than those of the Earth and of all the other planets. On its surface can be seen only a few and ill-defined light and dark patches. These patches do not seem to change their position and therefore we must conclude that the planet, which revolves around the Sun in 88 days, must always show the same face towards the Sun, just as the Moon does in relation to the Earth. Because of its small dimensions and the very high temperature of the hemisphere turned towards the Sun, which is between 350° and 450°C., Mercury must have been unable to retain the heavier gases. Mercury is probably composed of rocks of a volcanic nature, similar to those on the Moon, and is devoid of an atmosphere.

Next in our journey, we would meet Venus. With regard to the question of whether life is possible or not on this planet, we have already seen (page 149) how difficult it is to reach any conclusions on account of the dense layer of cloud which surrounds it. Already, however, the exploration of this planet has started by means of Mariner II and no doubt will be carried even further by means of spacecraft.

Beyond the Earth we find Mars, the planet which has been so actively studied and has been a source of so many discussions. We have already described (page 158) the extent of our present knowledge and the preliminary results obtained by Mariner IV.

The giant planets of our solar system, namely Jupiter and Saturn, are without any doubt liquid and gaseous, as we can deduce from the appearance of the peculiarities of their surfaces. Both these planets have a period of revolution around the Sun which is much longer than that of the Earth and at the same time they rotate with a fairly high velocity, their period of rotation being of the order of 10 hours. As a result of this, both planets appear considerably flattened at the poles. Since they are so far from the Sun, their surface temperatures are extremely low, lower than $-100°C$. If water existed on them it would be in the form of ice, while there is strong evidence to show that their external envelopes are composed of ammonia and methane. The mean density of Jupiter suggests that in the interior there must be a large rocky nucleus, while the mean density of Saturn suggests that its nucleus is much smaller and is surrounded mainly by liquid hydrogen and helium and by ammonia and methane in lesser quantities.

Uranus, Neptune and Pluto, the furthest planets from the Sun, have surface temperatures which are even lower. Uranus and Neptune have an atmosphere composed mainly of methane while Pluto is very small and is probably a solid body having a temperature which is very near absolute zero.

From this rather incomplete knowledge of the planets of the solar system we are led back to ancient beliefs that the Earth is the only planet privileged to be the abode of *homo sapiens*. Recent discoveries, however, of invisible planets belonging to other solar systems, must remind us that there are other infinite possibilities of Creation.

Translator's note:

The period of rotation of Mercury has been accepted in the past as being of 88 days. Recently, radar observations and a study of photographs and drawings of the surface of the planet taken over a period of years, indicate that the period of rotation of Mercury is of 58·6 days.

11 · From Halley's Comet to Arend-Roland's Comet

In some chronicles of the year 1456 there are descriptions of the apparition of a great and terrible comet of huge dimensions with an extremely long tail stretching for 60°, namely a third of the visible sky. Among his other official duties to the city of Florence, Paolo dal Pozzo Toscanelli also had to deal with political astrology in which he did not believe. He observed the comet accurately in its rapid movement across the sky. This was the comet which, according to a legend, was excommunicated during its apparition by Pope Calixtus III, so as to prevent all the calamities that such a portent would bring to humanity. In the search for the origin and for the truth of this legend, Father Stein, the late director of the Vatican Observatory, found an original papal bull of that period in which Pope Calixtus ordered solemn processions and prayers and church bells to be rung at midday in order to implore the help of God against the Turks. The almost simultaneous promulgation of the bull, the setting out of the first procession in Rome and the apparition of the comet at its brightest, which the astrologers of the time considered as a very unfavourable omen for humanity, led Father Stein to conclude that this was the reason for this extraordinary legend.

Halley, the second Astronomer Royal, in his investigations on comets at Greenwich, discovered that the bright comet of 1456 appeared again in the following years at least three times. The third occasion, in 1682, was the one that he himself could observe. According to his calculations he deduced that this was a periodic comet which describes an orbit around the Sun with a very high eccentricity. In its motion it passes relatively near to the Earth and we can observe it in its full splendour every 75 years. The last apparition of Halley's comet took place in 1910 so that the next one is due in 1985.

Can we be sure that the comet on this future occasion will be as bright as it was in its previous appearances which can be traced back through time to many centuries B.C.? While astronomers can, nowadays, predict with accuracy the return of a comet, it is difficult to assert whether its brightness and its general appearance from head to tail will be the same as before or whether during its long journey through the solar system the comet has undergone changes. It is certain that other comets have suffered changes. We can quote as an example the comet discovered in 1826 by Biela, an Austrian officer and an astronomer. The Biela comet, as this comet is known, has a period of $6\frac{1}{2}$ years and was traced back as far as 1772 and was observed until the year 1846. In that year, under the very eyes of the observers, the comet split into two parts which continued their motion side by side at a distance of 155,000 miles, for more than three months. Each part developed its own nucleus and its own tail which were seen again in 1852, although on this occasion the distance between the two parts had increased considerably. Since then the comet has never been seen again.

Meanwhile Schiaparelli announced his remarkable discovery that shooting stars or meteors, are the product of the disintegration of comets. They consist of minute particles which the comets have left behind in their orbit around the Sun on account of the disintegrating force exerted by the Sun and the planets upon the very rarefied matter of which the comets are composed. In the years 1872 and 1885 instead of seeing again the two parts which originally formed Biela's comet, observers from the Earth witnessed two magnificent meteor showers. This confirmed Schiaparelli's discovery. Nowadays we know the link between a dozen comets and the meteor showers which they have produced.

183

The comet 1956h (Plate 18), discovered on November 9th, 1956, at Uccle Observatory by the astronomers Arend and Roland, had great publicity because the calculation of its orbit around the Sun indicated that it would approach the Earth. The comet, which at the time of discovery was in Cetus, appeared through telescopes as a nebulous little star of about sixth magnitude and was followed by a tiny tail. It was rapidly approaching both the Sun and the Earth and therefore was bound to increase in brightness.

Among the many mysteries of the universe which man tries to unravel, we can include that of the origin of comets. Spectroscopic observations have given us considerable information about the physical constitution of comets. Thus we know that they are composed of the same elements as the stars, the Sun and the Earth. In the nucleus of the comet there must exist the heavier elements, namely metals or stony compounds like those we find in meteorites. The 'coma', surrounding the nucleus of the comet, and the tail are composed of molecules of hydrogen, carbon, oxygen and nitrogen, namely the very same gases which in various proportions and combinations form stars and planets. But why should these celestial bodies be so different and so strange compared with the others? This may well depend upon their mysterious origin and the great influence that the Sun exerts upon them. This influence is seen both in the attraction which maintains comets in their orbit around the Sun and in the development and orientation of their tails due to the pressure of solar radiation.

Several theories can be advanced. We can assume that comets were formed when the solar system itself was formed, because it seems highly probable that comets belong to the solar system. On the other hand we can assume that comets have their origin in the great disturbances which occur in the Sun. Large quantities of matter ejected from the Sun could well condense into masses relatively small such as comets which would remain subject to the attraction of the Sun. Future observations and theories may, in due course, either support or refute these hypotheses. Meanwhile our present knowledge should suffice to dispel any superstition which may still exist concerning the appearance of comets.

Many questions were put to astronomers about the Arend–Roland comet when it was not seen in the sky for a few days. During March and the first half of April 1957, the comet could not be seen because it was approaching the Sun. Calculations showed that by April 18th, when its motion would have taken it away from the direction of the Sun, the comet would, once again, be visible in the evening sky in a north-westerly direction.

In general, comets are nebulous objects and very diffuse, hence, with the exception of some special cases like that of Halley's comet in 1910, their brightness is easily lost if the sky is not very dark, which can happen not only at the time of twilight but also where the glare of the lights of a town reduces the darkness of the sky. Astronomers predicted that the brightness of the Arend–Roland comet would be comparable to that of the brightest stars of the northern hemisphere, and that it would have a tail already decreasing in brightness as the comet moved rapidly away from the Earth. In Italy the weather was not very favourable at the predicted time. In addition the haze near the horizon prevented all but a few observers from seeing the nucleus of the comet after sunset.

In Elba, good weather and a clear horizon enabled people to see the comet at sunset at about 8 p.m. on April 22nd, 1957. It appeared against the background of the sky which was still fairly light, and below the brighter stars of Perseus, which were just visible in the twilight. The tail appeared very diffuse and its length was estimated to be 5°. It is useful to remember that the disc of the full moon measures half a degree and that therefore the tail of the comet was almost as long as about ten moons placed side by side. On the evening of April 23rd, the comet, as seen from Elba, appeared at its maximum brightness since by then it was further away from the Sun and could therefore be seen later in the evening against the background of a darker sky. The very brilliant nucleus was surrounded by a coma and was followed by a long tail which could be seen to extend 15° reaching the stars of Perseus. Some observers claimed that the tail, which was directed towards Cassiopeia and was opening up in the shape of a fan, appeared to wind in a great spiral. Probably this was due to the fact that the comet actually had two well-defined tails, a phenomenon which had been observed two days earlier and which was clearly visible in photographs obtained at various observatories. Similar phenomena had in the past been observed in the case of other comets, indeed occasionally observers had witnessed the disintegration of the tail or of the nucleus itself. On the night of April 24th, the sky in Elba was not as clear as on previous nights, nevertheless the comet could still be seen, displaced further to the north, with its tail perhaps even longer and still divided into two parts at the extremity furthest from the nucleus.

After this, as the comet was moving rapidly away from the Earth, it could gradually only be seen through a telescope. From the nature of its motion we can say that the Arend–Roland comet, unlike periodic comets, will never again be visible from the Earth.

It is very unlikely that the Earth could meet the nucleus of a comet, but if this were to happen, probably all we would see would be a wonderful display of meteor showers. On the other hand the Earth has, in the past, travelled through the tail of a comet without any special effect being recorded. We can explain what the tail of a comet actually is and why it is harmless to human beings, by making a simple comparison. Let us imagine a beam of sunlight in a darkened church or the beam of a searchlight which breaks the darkness of the night sky. The dust particles which are always in motion either in the interior of a church or in the free atmosphere and are illuminated by the Sun or by the searchlight, seem to produce a real beam of bright particles which are in continuous agitation and which are extremely transparent. The tail of a comet is something of a similar nature although we must admit that the gases of which it is composed, namely hydrogen, carbon and nitrogen, are obnoxious to human beings. In other words, the tail of a comet is nothing more than a very transparent plume of smoke. This can easily be proved by the fact that stars seen through the tail of a comet do not lose any of their normal brightness.

The next question which may be asked is how the tail is produced and why it always appears to be pointing away from the Sun. The explanation is fairly simple if we consider the fact that very light gases are released from the nucleus by the heat of the Sun, particularly when the comet is approaching it, and are repelled by the pressure exerted by the light of the Sun. Light is a form of energy and more intense light has also higher energy. The light of the Sun on reaching solid objects exerts a pressure. In the case of the Earth it can be shown that the pressure exerted by the light of the Sun on the whole of the Earth is equal to 20,000 tons. Although this pressure is considerable, it is smaller than the powerful gravitational attraction of the Sun which keeps the Earth in its orbit. If the Earth, instead of being solid and massive, was as light as a leaf, the pressure of the light from the Sun would push it away in space just like leaves are blown by the wind. To sum up, all this may explain the origin of the tail of a comet which is composed of very light molecules of gas released by the nucleus and pushed by the Sun into space in a direction away from it. Drawings and photographs of comets observed in the past, from ancient times to the present day, always show this very definite characteristic of the tail which, however long or bright, always points away from the Sun.

There have been some exceptional cases when branches of the main tail appeared at right angles to the line joining the comet to the Sun, or indeed even directed towards the Sun itself. The comet of 1882 showed this peculiarity. In October 1882, Riccò at Catania and Tempel at Arcetri, remarked in their observations that for a few days the comet had 'a well-defined tail, a few degrees long and directed towards the Sun'. The Arend–Roland comet showed a much more exceptional and remarkable feature, which had never before been recorded. After passing nearest to the Earth on April 21st, 1957, and while receding from the Sun it was observed both visually and photographically between April 24th and 26th. These observations showed that the comet had a bright spear pointing to the Sun, namely in an opposite direction to its great fan-shaped tail. The comet had the appearance almost of a swordfish (Plate 18).

Two theories have been put forward to explain this exceptional phenomenon. It may have been due to the presence of heavy particles which are not subject to the pressure of light, but rather to an explosive action and to the attraction from the Sun. This hypothesis seems improbable because a phenomenon of explosive nature would not have shown itself in this way. The second theory is more acceptable. This suggested that it was simply a question of a second tail divided from the main one which was opposite to the Sun and appeared to point towards the Sun only because of an effect of perspective. The hypothesis seems to a certain extent to be confirmed by the fact that during the days when the spear was visible, it gradually acquired a definite tilt with reference to a line joining the Earth to the Sun. In other words, the appearance and the direction of the second tail were dependent on the position of the three bodies, namely the Sun, the Earth and the comet. Since the orbit of the comet had a large inclination to that of the Earth, in reality the second tail, seen in profile, was turned towards the Earth and not towards the Sun.

In any case we must admit that the Arend–Roland comet presented us with a strange and very unusual phenomenon.

12 · Cosmic Rain

Since the beginning of his existence on Earth, man has known at least two types of beneficial rain. First the rain of luminous rays or photons, as physicists call them, which reach us in great quantities from the Sun, the Moon, the planets, the stars and the nebulae. The other is the more common rain which comes from the clouds and which can be disastrous rather than beneficial if it is excessive.

Nearly half a century ago physicists discovered yet another type of rain which is invisible and is still a mystery, namely the cosmic rays. This type of rain has been widely discussed but only in recent years has it been possible to begin to understand its nature, its origin and the manner in which it continuously enters the terrestrial atmosphere. This knowledge has been acquired by investigations carried out in all parts of the world. Observations have been made underground, under water, at various latitudes and heights by means of balloons, aeroplanes, rockets and with extremely delicate and sophisticated new instruments.

The desire to learn much more about this particular pheno-menon has led scientists to advance theories about its origin. At a meeting of physicists and astrophysicists held at Climax in Colorado, where there exist special stations at 11,483 feet above sea level for the study of the Sun and of cosmic rays, Fermi put forward a theory on the origin of cosmic rays which seems to be increasingly confirmed by experimental tests.

It has been observed for some time that, as in the case of radio waves, some of the cosmic rays undoubtedly reach us from interstellar space. About 50 years ago, physicists, by using instruments which detected electric charges, noticed that such instruments were always affected by mysterious charges of unknown origin, no matter where and how their experiments were carried out. Whether on the frozen Canadian lakes or on the Eiffel Tower or on balloons as high as three miles above sea level, this mysterious rain affected their instruments. Moreover, physicists discovered that the intensity of this rain increased with height, which led them to suspect the existence of a penetrating radiation coming from above and most likely having an extraterrestrial origin.

World War I interrupted the investigations but soon after the end of the war, Millikan, Nobel prize winner, and Bowen, who became director of the Mt. Wilson and Mt. Palomar observatories, sent up balloons which carried recording instruments into the upper atmosphere to a height of nine miles. These experiments confirmed the increase of both the number and the intensity of electric charges present at this height and since then we have begun to speak of 'cosmic rays'. But what are these rays? Where do they originate? How can they penetrate the terrestrial atmosphere, water and indeed even the solid crust of the Earth?

Naturally the investigations were intensified and various types of instruments showed the nature of these rays, namely whether they were waves or actual particles travelling through space. To the very simple and elementary instrument known as an electroscope, which reveals the presence of electric charges, was added the 'ionization chamber', a type of instrument widely used nowadays to detect the presence of radioactive minerals and which is capable of counting the number of rays reaching it. Then the 'cloud chamber', capable of photographing the number and the trajectory of the rays which go through it, was introduced. Finally photographic plates were used with special emulsions which are affected by the rays which hit them so that the path and the intensity of the rays can be revealed.

Rossi had predicted that the positive or negative sign of the charge of the particles could be determined from the direction they hit the Earth with reference to the meridian of longitude and the parallel of latitude. We already know that the Earth behaves like a huge magnet, with magnetic poles which do not coincide with the poles of rotation.

Following the laws of magnetism, if a charged particle approaches the Earth in the neighbourhood of the equator, it will be deflected to the east if it is positively charged, and towards the west if it is negatively charged. With arrays of ionization chambers pointed to the sky like ordinary telescopes, it has been possible to show that the primary rays in their race towards the magnetic field of the Earth, are particles positively charged and endowed with extremely high energy. These particles reach the whole of the Earth at the rate of billions for every second. At the latitude of Turin, 3,000 to 4,000 particles per second per square metre reach the higher parts of the terrestrial atmosphere. Since hydrogen is the most abundant element in the universe, these particles, which have such a high energy and frequency, must be protons, namely the nuclei of hydrogen atoms. This element which is to be found everywhere and which carries within itself such an extraordinary power, does no harm to the human race in this particular case, because the cosmic rain, although so intense and of such high energy, is innocuous. This is due to the fact that it is widely scattered over the whole of the Earth and also because the mass of the particles of which it is composed, is extremely small and moves with a velocity which approaches that of light.

Photographic plates exposed at heights of about 19 miles, show that some particles leave traces which are so intense as to lead us to the conclusion that in addition to protons there must also be present particles of elements heavier than hydrogen, as for instance sodium or iron. Here we find an important link between the physics of cosmic rays and astrophysics. By means of spectroscopy, astrophysicists have revealed the presence in space of enormous masses of interstellar gases of low density, consisting of hydrogen, sodium, iron and other elements.

But let us return to the cosmic rays. Rossi made a considerable contribution towards the explanation of the manner in which these rays hit the Earth. By using lead sheets in the instruments detecting cosmic rays, he and other investigators showed the difference which exists between a primary and a secondary radiation. It is extremely rare that a proton can travel through the whole of our atmosphere, because it collides with the atoms of oxygen, of nitrogen and of the other gases present in the atmosphere. In the collision the proton loses some energy and produces a great number of secondary particles, each one having a mass which is smaller than that of the proton. When the surviving primary rays, together with the secondary rays, proceed towards the ground they penetrate layers of the atmosphere which are of increasing density. As a result of this the frequency of collisions increases and the particles continue to lose energy rapidly. At a height of approximately three miles, only a relatively small number of particles have enough energy left to be revealed by our instruments, and at sea level there are even fewer particles left.

It is only right to say that we are still very much in the dark as far as the origin of these particles is concerned. Nevertheless, theoreticians are always ready to put forward some new theory, and this is not a waste of time since it enables the experimental work to progress. We know that magnetic fields exist on the Earth, on the Sun, in the stars and probably in the galaxies as well. It was Fermi's view that the particles which constitute the cosmic rain can be accelerated by repeated collisions with magnetic fields existing in space, as happens in the powerful machines built to split the atom. The process appears logical enough. The particles of the cosmic rays would increase their energy by means of a kind of chain reaction following successive collisions. Fermi was able to prove his theory in the case of protons, but he himself recognized that the theory was not satisfactory in the case of heavier nuclei.

The fascinating problem of cosmic rays continues to be studied very intensively. Even if we do not reach a complete solution which may well be linked to the origin of the universe, we shall certainly witness some great and valuable progress in this field in the near future.

13 · Shooting Stars

The study of shooting stars, or meteors, has been of great interest to humanity since the early days of astronomy when it was believed that in effect stars detached themselves from the canopy of the sky to fall on the Earth. In recent years two new methods of observation have led to considerable progress in the investigations concerning meteors. At first sporadic meteors and meteor showers could only be observed visually and the observations made by Father Secchi and by Schiaparelli are well known. Later photographic methods were used. Radar astronomy and space probes have more recently widened considerably the study of meteors.

Besides the question of determining the physical nature of these bodies which enter our atmosphere and burn up, there are many other relevant questions to be answered. These concern the velocity and direction of meteors, their origin and moreover the number of meteors the Earth meets in its orbit around the Sun and in its revolution around the centre of the Galaxy as part of the solar system. Generally speaking, the greater the celestial body the greater is the quantity of matter which it contains, or in a single word what is called its mass. The Sun, which has a mass 330,000 times that of the Earth, exerts a gravitational force which is nearly 28 times that of the Earth. Hence we can easily imagine how many more meteors must fall daily on the Sun in comparison with the number which falls on the Earth. On the other hand the Moon, which is so much smaller than the Earth and the Sun, will attract fewer meteors. Because of the absence of atmosphere around the Moon, meteors cannot burn in their fall towards its surface but it is strange that no one has ever observed the actual fall of a large meteor upon the face of the Moon. Perhaps this is due to the fact that these bodies are so small that they can escape detection even by means of large telescopes.

In order to determine the quantity of matter that reaches the Earth either by landing on it or by burning itself in our atmosphere, it is necessary to have some indication of the size and weight of meteors. Let us remember that as these are the products of disintegrating comets, they are rather small bodies, being fragments of rocks or of metals. Because of the great velocity with which they travel through our atmosphere, they are ignited by friction and as a consequence burn and vaporize. Some fragments are larger than others and when they ignite they appear very bright in the sky, reaching a brightness comparable to some of the stars, but many show very little brightness. Occasionally the fragment is of a considerable size and in such cases it lands on the Earth. These are the meteorites which may weigh several tons and which can be studied in our terrestrial laboratories.

Investigations first carried out visually and photographically and, in more recent years by radar, have enabled us to determine the dimension of meteors and approximately the total number that the Earth meets daily. By means of radar we can record the echo from the traces left in the sky when meteors burn in our atmosphere. To this new method of observation has now been added the artificial satellite. This in its orbit around the Earth collides with meteors and the special instruments it carries are able to record such collisions and transmit their number to Earth.

It has been calculated that a meteor which when burning in the atmosphere appears as bright as Sirius, weighs probably between 1 and 2 grams. If, on the other hand, the brightness is only comparable to that of a star just visible to the naked eye—a sixth magnitude star—then the weight is only of the order of 1 milligram. It is easy from this to deduce how many of these meteors are necessary to make up the considerable weight of about 1,000 tons which is the estimated weight of matter received daily by the Earth. No doubt the data which we can now collect by means of artificial satellites will greatly increase our knowledge of this subject. At any rate, even if the total mass of meteors and meteorites were greater than the present estimate, it is certain that we are dealing here with a long term phenomenon in the sense that thousands of years will be required before a measurable change in the weight of the Earth can be detected. Let us not forget that the weight of the Earth is 5,977 million million million tons.

In addition we must remember that the majority of meteors are transformed into gases, which remain in the high levels of our atmosphere. Artificial satellites have shown that the atmosphere of the Earth, even if in a very rarefied condition, extends further than we used to think. What happens to this rarefied atmosphere which becomes diluted into space? Until recently we believed space to be empty but we now know that it contains dust made up of many elements. If on the one hand the Earth acquires weight does it not, on the other hand, lose some?

Whatever the answer is to these questions it is certain that the increase of weight so far determined is far too small to produce measurable changes both in the period of revolution and of rotation of the Earth.

Part III   THE STARS

1 · Man Reaches Out into Space

The exploration of space made possible by rapid progress in the construction and launching of space probes gives us an increasing knowledge of the universe in which we live. It seems to us more logical to use in this context the word 'exploration' rather than 'conquest', particularly if we consider how immense is the space which surrounds us and that man will probably never reach its limits.

Now that we have already obtained some successful results from the exploration of interplanetary space it seems salutary to consider whether it will be possible to make further progress. We have already obtained from our terrestrial observations a good knowledge of the bodies which form the solar system. We are certain that they, like all other celestial bodies, are composed of the same elements we know on the Earth. In addition we also know the temperature prevailing on the planets as well as many of their other physical characteristics. All the information available indicates that it would be very difficult to find on the other planets of the solar system more pleasant conditions than those existing on the Earth.

The conditions which we imagine exist on the other planets and their satellites have often been described. One of the great mysteries appears to be that concerning Venus, which is always covered by a very dense shroud of clouds, so that, even if on the planet there existed reasonable conditions to support life, the inhabitants would never be able to enjoy the beauty of a starry sky. It is true that this is only an hypothesis which only the direct exploration of the planet would in due course confirm.

One of the serious difficulties facing the potential explorers of space are the great distances involved. It is easy to express such distances in miles, but they would appear just as a string of noughts which really would not convey much information to the human mind. Perhaps the whole question could be better grasped if we expressed distances in terms of the time taken to cover them at a speed of the order of that reached by spacecraft. Let us therefore take the speed which we have to impart to a spacecraft in order that it could escape from the solar system, or in other words escape from the gravitational pull of the Sun. The calculated velocity would be 25 miles per second, equal to 90,000 miles per hour. With this speed, the time required to reach the Moon is 3 hours; to reach Mars 16 days; to reach Pluto, the outpost of the solar system, 5 years. To reach the nearest star to us in Centaurus, it would take us 30,000 years, and to reach the Andromeda Nebula it would take us 15,000 million years! So it is clear that once we have left the solar system in order to visit the stars, the time or the distance increases rapidly.

However, there is another reason why a spacecraft would not be able to reach a star, any more than the Sun. As already mentioned, the temperature of the Sun, which is not one of the hottest stars, is about 6,000°C., hence its radiation is so intense that even at some distance from it, a spacecraft would be completely vaporized.

During the International Geophysical Year and in the following years, a great many problems, some of which we have already discussed, have been studied and solved by means of artificial satellites which can travel in the terrestrial atmosphere and beyond. Instruments carried by these satellites can feed a great deal of data back to the Earth and this information has been of great interest to geophysicists as well as to astrophysicists.

In that field of astrophysics which deals with the exploration of the universe, it is very important, as we have already explained earlier, to obtain observations from outside the terrestrial atmosphere in order to eliminate its absorption.

In this particular field the problem to be solved is that of making spacecraft capable of carrying heavy equipment which can take observations automatically. We have already discussed in previous chapters the experiments which have been carried out and the results which have been obtained from the observations of the Sun, of Venus, of Mars and of the Moon.

The technique of launching manned spacecraft is rapidly improving as has been shown by the Gemini series of launchings in America. The question of more powerful propellents is being studied continuously. Careful thought is being given to a possible atomic engine, but there are still many difficulties to be surmounted.

Radio communications from great distances was another question which, for a time, worried the technicians, but the excellent results obtained during the Mariner IV mission to Mars have dispelled many doubts which existed on the power of transmitters.

There is no doubt that the exploration of space, although fraught with many difficulties, has a fascinating future. New knowledge of great scientific importance is already accepted not only in geophysics but also concerning the planets nearest to us.

Who can possibly foretell what the future observations may reveal to us and the importance of such knowledge? This by itself is a justification of the enormous effort, both intellectual and financial, which humanity is making at present.

One thing is certain, the exploration of space has definitely entered into the domain of scientific research and in this domain it certainly represents an investment which will be of great value to humanity as a whole.

2 · The Life of a Star

Light travels with great speed through space for millions and millions of miles. It brings us information about the stars which populate the universe at various distances, by means of radiations to which our eyes or our photographic plates are sensitive.

Among the large family of stars in the universe there exists a great variety. Nevertheless, stars can be divided into three large classes for which it has not been very difficult to find a logical explanation. Sirius, the brightest star in Canis Major, Vega in Lyra, Altair in Aquila and many other stars appear to be whitish-blue in colour. Arcturus, which guards the seven stars of Ursa Major, Capella in Auriga, Aldebaran, the eye of the Bull, belong to a class of yellow stars to which our Sun belongs. Finally, there are many stars like Antares, the brightest star in Scorpio, Betelgeuse, the shoulder of the giant Orion, and the brightest star in Hercules which are all red in colour. These three colours and all their gradations are even more conspicuous if we observe some double stars through a telescope. Albireo, the second star in Cygnus, is a double star of which the brighter is yellow and the fainter companion is blue. Rigel, the white star in the leg of Orion, has a faint companion which is very blue, and finally we can mention also a double star, Bootes, of which one of the two components is yellowish-red and the other white.

Early students of the physics of the stars, among whom we can name Father Secchi and Father Sestini, followed a simple line of reasoning. The blacksmith who is heating an iron in his workshop knows that at first it becomes dark red in colour, then red, orange, yellow and finally white. This is more or less the scale of colours which we observe in three classes of stars. Naturally astronomers immediately thought of a scale of temperature. The Sun compared with the other stars is fairly near to us and it was not difficult to determine its surface temperature and following the progress of spectral analysis it became possible to determine also the temperature of the stars, from the very dark red to the white ones. The scale of temperature of the stars ranges from 2,000° to 30,000°C.

At this point we may begin to think that these celestial bodies have a beginning and an end, or what we would generally call a life. We must, however, be quite clear what we mean by life. On the Earth we are accustomed to animal and vegetable forms of life, which once begun will continue as long as the conditions for their development are favourable. We generally say that the Moon is a dead world, because from terrestrial observations we see no changes in its surface, and we know that there is no atmosphere there, hence the Moon lacks those conditions which we consider essential for life as we understand it. The absence of life may only be relative, because even if in the interior of the Moon there are no gases such as methane and other hydro-carbons which we have on the Earth, nevertheless some of the rocks of which the Moon is composed, could contain uranium or some radioactive elements. These follow an evolution cycle of their own during which energy is emitted and in some way and at some future date, could develop some form of life. Celestial life is naturally very different from terrestrial life since it had a be-ginning in a remote past with the formation of new celestial bodies. These bodies certainly follow an evolution and individu-ally will reach an end, even if the matter which is transformed will not be able to start a new life.

The picture of the life of a star might appear at first simple enough since it is a question of considering extremely hot stars whose mass gradually changes into energy, passing through de-creasing degrees of temperature. In reality we discover that stellar evolution proves to be very complicated indeed, as our knowledge of the physical conditions of the stars increases. This is not surprising since stellar evolution is naturally linked to the creation of the universe and to the quantity of matter and energy it contains which is so great that it is beyond our comprehension. For stars whose distance we can determine, we can easily obtain their absolute magnitude by measuring their apparent brightness. Thus we can compare the brightness with the real dimension of these stars. As a result of this we now classify stars as 'dwarfs', that is to say of dimensions relatively small compared with the next group of 'giants' and the 'supergiants' which are of even greater dimensions. There seems to be a lack of continuity in the development of these three classes. Both in our Galaxy and in the galaxies, which are similar systems, we have red giants, blue supergiants and yellow and red dwarf stars all of which follow some sort of distribution. So far we do not know which is the beginning and which is the end of this distribution.

The Sun is a dwarf star and so are the majority of the stars nearest to us. Supergiants and giants have very extensive gaseous atmospheres which are very rarefied and hence have diameters which are hundreds of times greater than that of the Sun. We are now almost certain that the energy in the stars which are so rich in hydrogen is the result of the transmutation of hydrogen into helium. Stars may be born with various dimensions and with different amounts of hydrogen and of other elements. A star can live without a cataclysm taking place as long as there is an equilibrium between the transmutation of hydrogen and the gravitational force of its mass. If its equilibrium is upset then the star may become a variable, as happens in the case of many stars, or it may explode as in the case of 'novae' and 'supernovae'.

The conditions of instability may be responsible for the apparent discontinuity which we observe in the general picture of the evolution or life of the stars irrespective of their composition, dimensions and colour.

This is a very fascinating problem and is as yet far from being solved.

3 · The Colour of Stars

In the wonderful spectacle offered to us by the starry sky, there is one thing missing and that is the beauty of all the colours and hues which we enjoy in nature from sunrise to sunset.

We may ask ourselves why should there be this absence of colours during the night since we know that all stars, whether smaller or larger than the Sun, have a colour. A careful study of the starry sky even with the naked eye, soon reveals the fact that stars are not simply bright white points of light on the dark background of the sky, but are coloured. We have already explained the reason for this. Among the brightest objects in the sky are the planets which, since they reflect the light of the Sun and in some cases have an atmosphere, appear yellow or red. In addition, there are a few stars which shine brilliantly, such as Sirius in Canis Major, Antares in Scorpio, Arcturus in Bootes, Rigel and Betelgeuse in Orion, which are blue, yellow or red. These are the only shades of colour which the human eye can see in the sky.

The scarcity of colours in the immensity of the night sky can easily be explained if we consider that the human eye gradually loses the faculty of distinguishing one colour from another as the light grows dim and fades into darkness. Hence, our eyes can only distinguish a difference in colour in the case of the brighter celestial objects such as the planets and some stars, although all the fainter celestial objects to the limit of our vision, are also coloured. Thus in the observation of the night sky we have an almost total apparent absence of colour. The introduction of photography, which nowadays has become one of the most important tools of astronomical observations, brought no change. Indeed, if anything, the situation became worse. Until not long ago the sensitivity of the photographic emulsions was limited to the blue and violet regions of the spectrum. As a result of this, blue stars of different brightness could be recorded on photographic plates while the red stars would not show. In this way the picture of the sky was obtained in black and white, as if the sky had been seen through a violet coloured glass; that is to say showing only the blue stars which are the hottest. With the progress in the chemistry of photography, emulsions have been developed which are sensitive to all the colours of the visible spectrum and, in addition, also to the ultraviolet and to the infrared. In recent years a new photographic technique has been developed in which use is made of plates which are sensitive to the blue region of the spectrum, and a blue filter is added so that the photograph obtained, still a black and white one, shows us the sky in a blue light. If instead we use plates which are sensitive to the yellow or red, then we have photographs of what the sky looks like in these colours. However, the photograph is still black and white.

The above method has also been widely used at Mt. Palomar for the preparation of the great celestial atlas (Sky Atlas) which was recently completed with the help of the Schmidt telescope. This technique has led to important discoveries. In some regions of the sky, and particularly in the Milky Way and in its neighbourhood, photographs taken in the various colours as explained above, show that there are stars visible in one photograph and not in another. Masses of gases are distributed in the different colours, with various densities, indeed sometimes they are so dense as to hide all the other stars which might exist behind them.

Further progress in photography as it is well known, has produced types of emulsions which enable us to take coloured photographs. It is natural that astronomers should have attempted to use such material for celestial photography. In the case of the Sun, which emits a great amount of intense light, the problem is easier. Nevertheless, it has proved extremely difficult to reproduce the colours of the spectrum of the Sun with all the very fine gradations from violet to red. The wonderful spectacle of the solar spectrum which is very much like a rainbow, cannot be reproduced in its whole range of colours even by the most advanced type of emulsion for colour photography. The use of this type of emulsion for photographing the night sky is still more complicated because the relatively faint light of the stars requires long exposures and emulsions which are particularly sensitive. What is required, therefore, is a powerful telescope, that is to say a telescope of large aperture such as the 200-inch of Mt. Palomar. Experiments have been carried out with this instrument in recent years and the results appear to be very promising.

In colour photographs, stars appear in all the gradation of colours from blue to red according to their temperature. In addition, there are three types of celestial objects which when photographed in colour reveal important facts concerning stellar evolution.

Most people know the Orion Nebula which can just be seen with the naked eye in the sword of Orion. When this nebula is observed through a telescope it is one of the most beautiful objects in our sky. In its brightest part we can detect four stars which form a trapezium. These four stars and some others which are all of a very high temperature excite the whole of the gas, mostly hydrogen, surrounding them. This excitation causes the gas to become bright by fluorescence following a process which is similar to that which takes place in the ordinary fluorescent tubes widely used for illumination nowadays. The mass of gas which envelops these stars as well as a very wide region nearby, appears as a luminous purple cloud which, against the dark sky, gradually becomes a very definite violet colour. Dark clouds, that is clouds of gas which are not excited, appear here and there in the nebula. The nebula which covers a very wide region in Cygnus belongs to the same class of irregular nebulae. Its appearance is very much like a long thin piece of lace or net of luminous matter enriched by chains of stars. In this nebula we notice many colours and all the gradations from red and blue to white. It is extremely difficult to advance any theory on the origin and on the formation of this strange and beautiful nebula. However, in the brightest parts, that is to say in those which are more highly excited, we seem to witness the actual birth of new stars or see the remnants of the explosion of a nova.

The second type of object successfully photographed in colour, is that known as 'planetary nebula' although it has nothing to do with the planets. Here we have a star which in the past exploded and as a consequence of this explosion the gases of its more external layers have receded and continue to recede, while still surrounding the original star. While the central star is blue and is gradually approaching the end of its life, the gases which surround it shine in the various colours from red to blue.

Finally, in the third type we have colour photographs of the well-known spiral nebula, or galaxy, in Andromeda. The nucleus is reddish in colour and contains cooler stars which are decaying. In the spiral arms, which are mixed with dark matter, the colour of the excited gas and of the younger and hotter stars is increasingly blue. Probably stars continue to be born in the spiral arms of this very complex system.

4 · Double Stars

The distances between stars are very great indeed compared with our standard of distances. If we want to find the nearest star to us, apart from the Sun, we must travel in space for more than four years with the velocity of light, which is 186,000 miles per second. The nearest star to us is in Centaurus, which is a constellation visible from the southern hemisphere. Space, therefore, between the stars would appear to be empty but let us not forget that in this apparent vacuum, there are always present gases and elementary particles even at very low densities. The stars may have families of planets like our own solar system where the distances between individual planets are small when compared with the distances between stars. There exist, however, many systems of stars in which the distances between the components of the system are relatively small. The law of gravitation maintains the various components of the system bound to each other and they move around a common centre which is the centre of gravity of the system. We are referring to what astronomers call double or multiple stars which are so common in the sky and which show a great variety of types.

A double star whose components are of equal mass will have the centre of gravity of the system half way between the two, and the astronomer will be able to determine the period of revolution of the two components around the centre of gravity. The periods of revolution may range from a few hours to some millions of years and this may give us an insight into the particular conditions which gave rise to these systems and also to their evolution throughout the ages. Double or multiple stars are rather common but there exists a peculiar class of systems known as 'eclipsing binaries'. From the name itself it is clear that we are dealing here with systems consisting of two stars which eclipse each other in turn. This can only happen when the plane of their orbit in the sky has such an inclination that from the Earth we can see one star passing in front of the other as they move around the common centre of gravity.

The well-known star Algol, the second brightest star in Perseus, is such a star. In 1668 it was discovered that this star was a variable, but more than a century had to pass before the periodicity of the variation of its light could be established and interpreted as being produced by the reciprocal eclipse of two components. In the sky many variable stars exist, that is to say stars whose light increases and decreases rapidly. The variability of these stars is due to an intrinsic cause, namely their internal instability and therefore, they must not be confused with the eclipsing binaries.

Let us return to Algol, which is a system consisting of two stars. The brighter is white in colour and has a very high temperature (about 20,000°C.), the fainter star is smaller, yellowish-red in colour and has a temperature of about 7,000°C. The period of rotation of the two components around the centre of gravity of the system is a little less than three days. During this period we can observe the variation of light which occurs when the two components eclipse each other. Both have dimensions which are much greater than those of our Sun and the distance between their two centres is six million miles, namely 15 times less than the distance between the Earth and the Sun. It is extremely difficult to imagine the phenomena which must develop between the components which are composed of gases at such a high temperature and which are so relatively near to each other. Let us for a moment consider what happens in the system Earth–Moon, both solid and cold bodies and so much smaller than the components of a double star system. The Moon, in spite of its small dimension, is capable of exerting a considerable action upon the large liquid masses of the Earth and produces the tides with which we are so familiar. The tides which occur between the two components of Algol and of all other similar systems must be truly gigantic.

Periodic phenomena, such as those which occur in systems similar to Algol, have a relatively easy explanation, but there are other systems which are much more complex, particularly those which show in the variation of their spectra some anomalies which are not easily understood. The astronomer attempts to explain these by advancing various hypotheses. We shall mention two such systems whose constitution now seems to be fairly well understood. The second brightest star in Lyra (β Lyrae) shows variations of light with a period of about 13 days, which can easily be observed with the naked eye. The variation of light of this system is due to the reciprocal eclipse of the two components which present complex phenomena. The hypothesis which has been put forward is that both components are surrounded by an extended and very tenuous gaseous envelope, forming what can be called a 'contact binary system'. The two components must be a white star and a yellowish-red one. Theory suggests that a system of this type, with components of unequal dimensions, must be unstable and that a flow of matter between the two stars must occur. The presence of this flow of matter could well explain both the complications observed in the spectrum of the system and in the various characteristics of the variation of light. The transfer of matter from the first to the second component and vice versa, would tend to shorten the period.

Similar complex phenomena are also present in the system ϵ Aurigae. This particular system has been observed and studied by many astronomers. Following recent observations at the Merate Observatory, it has been possible to advance an hypothesis which seems to explain better than any other previously put forward, all the results of the observations of its spectrum and of its light variations. Every 27 years the ϵ Aurigae system goes through an eclipse which lasts two years. During a whole six months the star grows fainter, then it remains constant for

211

18 months at half its original brightness and finally, during the last six months, gradually returns to its normal brightness which it maintains for 25 years.

It seems that the system in this case also consists of a yellowish-red star having a temperature of 7,000°C., a mass 30 times that of the Sun but dimensions much smaller than the other component. This other component is a white star with a temperature of about 25,000°C., a mass 20 times that of the Sun and is surrounded by an extended envelope of hydrogen and other gases which are excited by the ultraviolet radiation which it emits. The envelope rotates together with the star which it envelopes and must be almost transparent because through it we can see the light of the yellowish-red component. Between the two components huge tides develop. From the length of the period we can deduce that the distance between the two components is much greater than that in the system of β Lyrae. It is about 13 times the distance between the Earth and the Sun.

5 · Exploding Stars

Nature at times seems to play tricks on man who is so anxious to investigate and to discover its mysteries. In the nuclear microcosm elementary particles seem to increase from day to day so that very much still remains to be explained about its essence. Similarly, in the macrocosm we seem to lack either the capacity or the possibility of understanding it in its immensity. In the end we must consider it to be infinite, both in its dimension and its vast quantity of energy.

Even in the progressive development of celestial phenomena limits of time do not seem to exist. While some phenomena develop with enormous speed so that we can follow their various phases, the majority of phenomena take an extremely long time to develop and it is very difficult for us to follow their evolution. We are unable to discover even any difference in structure depending on the age of galaxies which are so far away from us that light, in spite of its very high velocity, only reaches us after several thousands of millions of years.

Among the celestial objects which undergo very rapid changes we can include that class of stars which show an unstable equilibrium. It is true to say that all stars which are spheres of incandescent gases are far from calm considering their constitution and their very high temperature, but generally speaking they show a constant brightness. Thus Vega, Arcturus and many other well-known stars appear to us always of the same brightness. Many other stars, however, those the astronomers call 'variable', show considerable light variations which can easily be followed, in some cases with the naked eye, but in many cases certainly with the aid of the telescope.

Stable and unstable equilibrium is always a relative phenomenon. Even the Earth, which is a solid body, can hardly be thought completely stable if we think of the earthquakes which so often shake it, or if we think of volcanic activity. The Sun, which fortunately for us is in a state of reasonably stable equilibrium, breaks its relative calm every 11 years in order to unleash storms of various degrees of intensity.

Much more remarkable must be the phenomena which occur in variable stars and of which we can hardly imagine the magnitude. It is enough to say that the large phenomena which occur in the Sun and which show themselves in the form of sunspots, hardly alter the light that the Sun emits. In the sky we have a great number of types of variable stars if we consider the amplitude of the light variation and its duration, namely the period during which such variations occur which may be regular or irregular.

The star δ Cephei has some special characteristics and all stars belonging to this type are called 'Cepheids'. It was discovered that their light variation is due to a pulsation of the star which periodically expands and contracts. In addition it was noticed that stars of longer periods are also stars which are intrinsically brighter. We shall discuss these stars in greater detail later.

Up to this stage we have dealt with phenomena which are rapid but not extremely fast. In the case of the Cepheids the period may range from a few hours to a few hundred days but there exist variable stars in which irregular phenomena of an explosive nature develop very rapidly. These stars seem to undergo a real cataclysm and because of their sudden apparition are called 'novae' and 'supernovae'.

The phenomena which cause these stars to appear in the sky are almost certainly due to explosions of stars which already existed. The internal process of the production of energy is subject to a cataclysm which produces the eruption usually not of the whole star but of its more superficial parts. Generally a nova is not a new star but is a faint star which suddenly increases in brightness. In fact the faint star can be found on photographs which may by chance have been taken before the explosion. The study of the spectrum of the nova reveals that there is an envelope of gas which expands with a velocity of the order of a few thousand miles per second. This gaseous envelope consists mostly of well-known elements, such as hydrogen, helium, oxygen and metals. The size of the phenomenon can hardly be conceived when we consider that the star may, in a very short time indeed, reach a brightness equal to a hundred thousand times that of the Sun. After that a slow decrease in brightness takes place which may last for several months, with fluctuations of intensity until the star once again reaches its original magnitude.

Even greater are the phenomena which produce the supernovae. In this case also the appearance is very sudden, but the increase in brightness is much greater, being 1,000 times greater than that of the novae. At the time of the explosion, supernovae reach a brightness which is 100 million times that of the Sun. Novae which appear occasionally in our Galaxy or in the far distant galaxies are rather rare. Even more rare is the appearance of a supernova. Calculations indicate that one supernova appears in a particular galaxy every 400 years. In our own Galaxy there have been only two or three within known historical time.

When a supernova explodes in a galaxy it appears as the brightest single star of the system and indeed in some cases it may be even brighter than the system as a whole. The indications are that supernovae explode from young stars which are very hot, have a large mass and which are born in the spiral arms of galaxies.

As we have already said, nature always has new surprises in store for us. We believed that in the supernovae we witnessed one of the greatest celestial cataclysms but astronomers at Yale have recently discovered that one of the furthest galaxies which can be observed with our instruments and which emits radio waves, undergoes fluctuations in its light. In reconstructing the history of this galaxy by means of the many celestial photographs available at the Harvard Observatory, astronomers have shown that since the early observations in 1887 this galaxy is subject periodically to sporadic but intense variations of light. The most plausible hypothesis is that in this galaxy, which is nearly 2,000 million light-years away from us, an exceptional supernova had periodically appeared. It is calculated that such a supernova must be at least 10,000 times as bright as any of the supernovae so far known to us.

6 · Flare Stars

Variable stars, as their name implies, are stars which show light variations which can be more or less regular. A great variety of these stars exists. There are those whose variability is due to a physical cause, such as the instability of their production of energy. Others whose variability is due purely to an optical reason, such as the eclipsing binaries, where the light changes are due to periodic eclipses of two stars which revolve around their common centre of gravity.

Novae and supernovae, the exploding stars, can also be considered as variables, because after a period of relative calm and faint luminosity there follows a sudden and very intense increase in brightness which can be repeated without regularity. Actually, all the stars, including even the Sun, are, in a sense, variable. We know that the Sun is covered by spots during a cycle which lasts 11 years. If we were a few light-years away from the Sun, we would not be able to observe the sunspots because although they can be numerous they affect only limited regions of the Sun. In reality the Sun, and probably all the stars, are subject to light variations even if these are not observable. The causes of these light variations are inherent in the constitution, in the varying instability and in the development of the nuclear reactions which take place in the interior of the stars.

Another type of star which can be included in the class of variables, is that which is subject to rapid and irregular flares. The word 'flare' has been introduced by astronomers to denote a phenomenon which occurs on the Sun when it is in turmoil. In the regions where sunspots are present, a few points, in the umbra and penumbra, may suddenly become very bright for a few minutes and then disappear. These points, or in some cases even small areas, emit X rays and ultraviolet radiation of great intensity which produce disturbances in the ionosphere and in the terrestrial magnetic field. Solar flares are eruptions from the photosphere and the chromosphere. The phenomenon applies only to very limited regions of the Sun, so that its general luminosity is not really variable. In the case of some stars similar phenomena can affect a considerable part of their surface with the result that the stars appear to flare up suddenly.

About 20 years ago, American astronomers discovered some rather peculiar stars which were very faint and had a fairly low temperature. These stars, named red dwarfs, seemed to be subject to flares of very short duration which had the characteristics of solar flares. At present about a dozen flare stars are known but their number is increasing rapidly, particularly on account of the discoveries and the continuous observations carried out at the Asiago Observatory. In this particular class of stars there is a typical one in Cetus, known as UV Ceti, which is about nine light-years away from us and hence can be said to be relatively near. This star is a red dwarf having a diameter which is only seven-hundredths that of the Sun and it has flares which increase its brightness about 10 times. Observations of its spectrum indicate that the flares are due mostly to incandescent helium and to a rapid increase of ultraviolet light.

Meanwhile radio telescopes began their search for celestial objects which emit radio waves. Since the Sun emits intense radio waves, it could be predicted that stars also must emit radio waves of various intensities. The search, however, was not an easy one and it proved very difficult to identify sources of radio waves with bright stars, perhaps because these stars are very far from us and radio telescopes are not yet powerful enough to detect radio waves from individual stars.

On September 28th, 1958, Sir Bernard Lovell used the 250-foot radio telescope at Jodrell Bank to observe UV Ceti, and succeeded for the first time in recording a radio emission from a star other than the Sun. In this particular case there was no doubt that the optical object coincided with the radio source. Since then the observations with the Jodrell Bank radio telescope have increased rapidly and, at the same time, astronomers have also intensified their observations in the search for flare stars by means of photography and photoelectric devices.

As we have already mentioned, the smallest flares last only a few minutes, the more intense ones may last as long as half an hour and may be repeated at intervals of about ten hours. In the case of UV Ceti the interval is 35 hours. In order to intensify the observation of this type of star, a programme of photographic observations was evolved by means of telescopes situated all over the world. It was of the greatest interest to be able to record simultaneously the optical observation of a flare appearing on a star with the observations carried out by radio telescopes. Working at various frequencies it has been established that the maximum intensity of a flare observed by radio, occurs simultaneously with the maximum optical intensity, although in the radio observations the flare appears to be of longer duration.

It is calculated that the temperature of UV Ceti, as a consequence of the flare effect, reaches a very high level, about 100 times higher than that which occurs in the most intense solar flare.

7 · The Wonderful Cepheids

Among the great variety of stars which populate the sky, we have discussed the exploding stars and the flare stars. We would now like to discuss another peculiar type known as 'pulsating stars'. Astronomers call them 'Cepheids' from the type star discovered by Goodricke in 1784. This particular star, δ Cephei, is the fourth in order of brightness in Cepheus. It is of almost fourth magnitude and hence visible to the naked eye. Its brightness varies by almost one magnitude in about $5\frac{1}{2}$ days. During this interval of time the star begins to increase rapidly in brightness, reaching a maximum almost in a day. After this it begins to decrease in brightness in a little more than four days returning thus to its minimum, only to start its cycle all over again. During these variations the star changes a little in colour, from whitish-yellow at maximum to yellow when at minimum of its brightness. We know that this is due to a lowering of its temperature from about 7,000° to 5,000°C.

The study of its spectrum shows that its gases are in motion. At first this was interpreted as due to the motion of two stars so near to each other that could not be resolved by means of a telescope and which rotated one around the other. With time many variable stars were discovered which showed the same basic characteristics as δ Cephei, namely a rapid increase in brightness and a slower decrease, with periods which ranged from less than a day to about 100 days.

Because of the appearance of the spectrum of these stars,

astronomers at first considered the Cepheids to be double stars and began to calculate their orbits. Soon, however, they discovered that all the orbits calculated seemed to have the same orientation and moreover that the two hypothetical components would have to be so near to each other that they would be partly encased in each other. Shapley, who for many years was the director of the Harvard Observatory, suggested that the Cepheids must have been giant stars having volumes from 15,000 to 20,000 times that of the Sun; that they were subject to periodic pulsations so that their radii periodically increased and decreased in relation to the variation of their light.

The systematic photometric and spectroscopic study of the Cepheids led, in 1912, to another remarkable discovery which was to be so important to astronomy. Miss Leavitt, of Harvard Observatory, followed the light variations of a group of Cepheids which belonged to the Small Magellanic Cloud which together with the Large Magellanic Cloud were first seen by Magellanus, hence their names. Later it was discovered that these objects are not irregular nebulae belonging to our Galaxy, but rather galaxies whose spiral form we can just see.

The Small Magellanic Cloud is about 200,000 light-years distant from us, and we can assume that the Cepheids, which belong to it, are approximately at the same distance from us. Since the Cepheids observed by Miss Leavitt were of various periods and of various magnitudes, she concluded, assuming they all had the same distance from us, that there existed a well-defined relationship between the period and the brightness of these stars.

Observations of Cepheids in various regions of the sky, confirmed the fact that as the period increases so does the luminosity, while as the density decreases, the volume increases rapidly. The result of this is that a Cepheid with a period less than a day, is about 100 times brighter than our Sun. From this discovery there immediately followed a very important method for the determination of the distance of celestial objects which, like the galaxies, contain Cepheids. Once the period of a Cepheid is established, from the period-luminosity relation, we can quickly determine its distance up to the limit imposed by our telescopes.

In figure 10 are shown some typical Cepheids of various periods in order to illustrate both the period-luminosity relation and their real dimensions.

The theories which have been advanced to explain the pulsation of these stars are indeed complex. Briefly we can say that

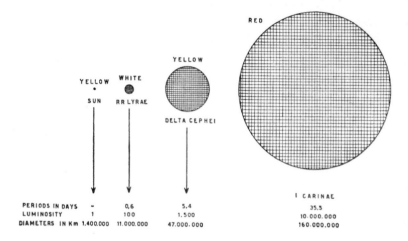

Fig. 10. Periods, luminosity and diameters of some Cepheids compared with the Sun

perhaps the phenomenon is not strange if we bear in mind the probable process which produces energy in the interior of a star. In a normal star it is assumed that at any point there is a balance between the weight of the external layers and that of the hot gases which constitute the nucleus of the star. If for any reason this balance does not exist in a star, because of an excessive development of sub-atomic energy, an increase of the pressure of the gases produces an expansion of the whole gaseous sphere. The limit of this expansion will be determined by the extent of the gravitational influence of the star, after which the internal pressure of the gas becomes too low to maintain the balance and the external layers of the envelope will start to fall back towards the centre of the star. This will happen with a rythmical period which in fact we observe in the Cepheids.

8 · Radio Stars

The radio spectrum of the stars, that is to say the emission of radiation of long wavelengths by the stars, is separated from the optical spectrum by a wide region of absorption mainly due to the water vapour in our atmosphere. On the side of the lower frequencies, the radio spectrum extends to the short waves and ends on account of the absorption of the ionosphere. Its limits can be fixed approximately at a wavelength of 100 metres on one side, and at a wavelength of $\frac{1}{2}$ centimetre on the other. If in this interval the radiated energy was purely of thermal origin, its distribution in the radio spectrum should be ruled by the laws relating to a black body. From experiments so far carried out in a short part of the spectrum, it appears that there is a preponderance of processes still unknown which are not of thermal origin, like those which occur in the solar corona and prominences.

Observations of the radio spectrum are carried out nowadays at many stations scattered all over the world, following various methods of reception. The radiation which we receive from some regions of the Galaxy, shows very small fluctuations which are not more than 10 to 15 per cent of the normal value. The solar radiation, on the other hand, occasionally reaches very high values, particularly when storms occur on the Sun. One of the first important observations was that made by Hey in 1942. The radiation which was very intense during the whole of the day of February 27th, disappeared completely at sunset only to reappear again the following day soon after sunrise. Obviously

223

it was a question of direct solar radiation and not, as it might have been thought, of a radiation produced in the ionosphere by the Sun. On that very day a group of sunspots with an area greater than 1,500 millionths of the sun's disc, crossed the central meridian of the Sun and very intense flares were observed at Greenwich, Arcetri and Sherborne. On February 28th a complete fade-out occurred on short waves, and 24 hours later a violent magnetic storm affected magnetometers. Hey estimated that the radiation, which was the cause of the radio noise, originated from a source with a temperature of 100 million degrees.

In recent years the Sun has been increasingly under observation by means of instruments specially designed for the purpose. Solar radio noise of permanent origin, that is emitted by the quiet Sun, is regularly recorded and so is that much more intense emission consisting of bursts which occur when the Sun is disturbed. Thus we can follow the relation which exists between the visible disturbances such as flares and sunspots, and the bursts which are recorded by radio.

The technical difficulties which are encountered in the methods of radio observations of the Sun and of the stars in general, are of various types. Each frequency requires specially designed circuits and aerials. Moreover, the energy received from the Sun and from the stars is normally so small that it is easily confused with the background noise of the receivers themselves, and the resolving power of the instruments used is rather small. The resolving power of a telescope, both for optical and radio frequencies, is directly proportional to the wavelength and inversely proportional to the aperture of the instrument. For a wavelength of 10 centimetres and an aperture of 10 metres, the resolving power is about one-hundredth of a radian, that is nearly half a degree. In order to reach a resolving power of the order of the human eye, which is about one minute of arc, it would be necessary to increase the aperture of the radio telescope to a very large extent. It is usual nowadays to use large parabolic reflectors which have an aerial placed at the focus. By combining a number of such reflectors to form an interferometer, it has been possible to reach a resolving power of the order of three or four minutes of arc. Even this resolving power is not sufficient to isolate a single star which may be emitting radio waves.

Radioastronomy aimed at detecting regions of the sky which emitted radio waves and then, within the limits of the resolving power of the instruments, tried to find out which celestial objects emitted the radiation. It is quite possible that stars, like our Sun, emit radio waves continuously, but because of their great distance from us the emissions may be too weak to be recorded by our instruments. When the object emitting radio waves consists of a considerable group of stars, like a cluster or a galaxy, or when it is a question of a nova or of a nebula in which phenomena of instability are present, then there is some likelihood that the radio waves emitted can be recorded.

Regions of the sky have been discovered where neither the eye nor the telescope nor photography reveal any celestial objects, and yet these regions are sources of radio emissions. Astronomy of the invisible is not new and we can foresee great developments thanks to the advent of radioastronomy.

The progress in the preparation of photographic emulsions has made it possible to extend the optical spectrum to the ultraviolet and to the infrared, thereby widening considerably the spectrum which used to be available for observations in earlier days. A whole new world of stars and nebulae, until recently unsuspected, has appeared in celestial photographs taken with plates coated with these new emulsions. Theory has also helped considerably in the discovery of invisible celestial objects. In the classical case of the discovery of the planet Neptune, theoretical predictions were so accurate that the planet was immediately found among the many stars of the sky. In other cases by the observation of irregularities in the motion of some star it was possible to discover the existence of bodies of small dimensions that are probably dark, which may, after all, be planets belonging to other solar systems. It is reasonable to predict that much more will be discovered in the astronomy of the invisible by means of radioastronomy. Only in a few cases do 'radio stars' coincide with stars and celestial objects which can be detected either visually or photographically. It could be suggested that radio stars represent a type of celestial object so far unknown, which is either completely dark or extremely faint and which emits radio waves with an intensity much greater than that of visible stars.

The Sun is the nearest star to us and therefore we can study it in greater detail than any of the other stars which appear to us only as luminous points. From the whole of the Sun there is emitted a background radiation which is constant in time. Moreover there appears to be a slowly varying component which is accompanied by exceptional bursts due to the presence of sunspots of moderate size. This component, which has a wavelength of the order of a centimetre, generally varies with a period of 27 days, which is the period of rotation of the Sun around its axis. Both the background radiation and the variable component can be explained by the temperature of the photosphere which, as we know, is of the order of 6,000°C. When the Sun is disturbed and large sunspots are present accompanied by flares and prominences then, to the radio emissions, are added bursts of varying intensity which can be classified in types according to the manner with which they reach the Earth.

The very high temperature of the solar disc, corresponding to the flux of energy emitted in these radio storms, cannot be explained as originating in thermal processes or having a common origin. Some bursts may be explained by oscillations of the solar plasma, that is by the ions, electrons and neutral electrical particles which are excited by a process which so far is of an unknown origin. Other types of violent bursts, which are associated with intense solar storms and which sometimes produce cosmic rays, can be explained by processes similar to those produced by a 'synchrotron' which is an instrument used in terrestrial laboratories for the study of atomic structure.

The discovery that the Earth acts as a huge magnet is a very old one. More recent is the discovery that the Sun also has a magnetic field which is well defined and which has an intensity a little greater than that of the terrestrial magnetic field, while magnetic fields associated with sunspots are much more intense. Another important discovery was made recently. Many stars have intense magnetic fields whose intensity would depend upon the physical conditions of the stars themselves. We now know that the Galaxy is a large stellar system, organized like the solar system but on a much larger scale. It has a centre around which revolve all the stars and all the gases of which both the bright and dark nebulae are composed. It was natural to think that the Galaxy also had a magnetic field which, incidentally, could explain the origin of cosmic rays according to Fermi's theory. The Jodrell Bank 250-foot radio telescope has recently discovered and measured the intensity of an extremely weak magnetic field in the Galaxy, with properties very similar to those of the magnetic fields of the Earth and of the stars.

Among the discovery of sources emitting intense radio waves we can mention three galactic supernovae. As we have already explained (page 214) these are stars which explode suddenly because of their internal instability. They are rather rare both in our own Galaxy and in the galaxies. In our Galaxy there have probably been three. The first can be identified with the phenomenon observed and recorded by the Chinese in A.D. 1054. The second and third were those observed by Tycho Brahe in 1572 and by Kepler and Galileo in 1604. In the position in the sky where these three supernovae were observed, we can still today see a faint nebulosity.

Another intense source of radio waves has been discovered exactly in the direction of the galactic centre in Sagittarius. The radio waves are probably produced both by a thermal and a nonthermal process.

The flare stars (see page 218) which periodically show a sudden increase of brightness, also emit radio waves similar to those emitted by the Sun. Finally we can say that soundings made in the depth of space by radio telescopes, have revealed that some galaxies, which are at a distance of several thousand millions of light-years, emit radio waves of various intensities. It is clear that because of their great distance, the phenomena which take place in them and which produce radio waves, must be on a very imposing scale of magnitude.

The main problem is always that of identifying radio sources with objects which can be detected optically. Only for about a dozen or so of these radio sources has it been possible with the help of the 200-inch reflector to make such an identification. In many cases the situation is complicated by the fact that radio telescopes are able to reach much further in space than optical instruments. The objects identified had spectra which were extremely difficult to interpret. At first it was suggested that they might be supernovae, but a careful study has shown that they are very remote galaxies. These galaxies, because of the general expansion of the universe, appear to recede from us with a very high velocity which increases in proportion to their distance from us.

Galaxies have been discovered which recede with a velocity of the order of one-third of the velocity of light. The brightness of these galaxies is much greater than that of our own Galaxy, but their dimensions are much smaller. It is thought that only gigantic explosions in their nuclei could possibly produce radio waves so powerful that they are still detectable by our instruments after a journey through space lasting a few thousand million years.

Part IV THE GALAXY, GALAXIES AND THE UNIVERSE

1 · Probing the Infinite

On board a ship, when the captain wants to know whether there is enough depth below the ship in order to navigate, he orders soundings to be taken which will give him the required information. To measure great depths electric methods are used nowadays. Signals are reflected from the sea bottom and in this way even the deepest oceans can be measured.

When we wish to probe the depth of celestial space, the situation is very different. The stars visible to the naked eye from both hemispheres of the Earth, are relatively few in number, only a few thousand, so that they can be counted. The ancient observers believed that the stars were scattered and embedded in a crystal sphere and hence were all at the same distance from us. With the progress of astronomy it was not long before it was realized that the stars were, in reality, at various distances from the Earth and that the brighter stars were the nearer and the fainter the further away from us.

With the introduction of the telescope, even in its modest size of Galileo's time, the number of visible stars was increased to such an extent that it became impossible to count them. Their number was increased even more as larger telescopes were introduced and astronomers began to ask themselves how to determine the distance of the stars from the Earth. This was far from being an easy problem with the rudimentary instruments which then existed. Much unsuccessful work was carried out in this direction before it became possible to measure the first distance of a star.

It was only just over a century ago that instruments became available which were so accurate that astronomers were able to measure very small angles, smaller than one second of arc. The first determination of distance was made for a star in Cygnus, which was one of the few stars nearest to the solar system. Soon it became evident that the direct determination, visual or photographic, of the small displacements of the stars in the sky, could only be obtained for the nearest stars, that is to say for stars which were not further than 300 light-years from us. These displacements are due to the apparent motion of the stars in the sky in the course of a year, as a result of the revolution of the Earth around the Sun.

The angles subtended by the diameters of stars are also very small, even if in reality these diameters are much greater than that of the Sun. In fact the angles are so small that only in exceptional cases and with special instruments, can they be measured, so that all the stars, even when observed with large telescopes, appear to us simply as bright points of light because of their great distance. On the other hand, stars of much smaller dimensions may appear very bright to us because of their relative nearness to the solar system. The apparent difference in brightness is due to the fact that we cannot see the diameter of the stars, but only the luminous energy emitted as a whole by the stars and which reaches us through space.

Several indirect methods, not all of them very satisfactory, have been evolved in order to determine the distance of stars which are beyond those whose distance can be measured directly. Thus, once the astronomers had obtained an indication of the distance of the various types of stars, they turned their attention to other important questions. They carried out star counts on photographs of several celestial regions and finally set themselves the difficult and lengthy task of probing the depths of space.

In spite of having available an instrument like the 200-inch reflector of Mt. Palomar which can reach a distance of a few thousand million light-years, it was known that observations would not reach the limit of space and speculations are rife on the question of the depth of space which to us seems to be boundless. A preliminary simple survey tells us that stars appear in greater numbers and seem to crowd around the luminous belt which we call the Milky Way. From these observations we can deduce that the Galaxy is, in reality, a very large stellar system, which, like the solar system, is kept together by the gravitation exerted by its mass and, in addition, that its shape is that of a spiral. The solar system, composed of the Earth and of the other planets, occupies a position which is far from the centre of the Galaxy . We can only see the Galaxy from the inside and therefore it is impossible for us to have a complete picture of it.

Fortunately for us there exists in the sky, as we have already mentioned, another stellar system or galaxy, which is visible from the northern hemisphere of the Earth. This galaxy shows several characteristics which are similar to those of our Galaxy but in this case we can see it as a whole, from the outside, projected on the sky not edgeways, but with an inclination which is sufficient to enable us to study it in detail. We are referring to that galaxy which is seen in Andromeda (Plate 3), visible to the naked eye but appearing as a hazy patch of light and which even when observed visually through a telescope, does not reveal much to us. Photographs of this object obtained with our great modern reflectors and with photographic plates which are sensitive to the various regions of the spectrum from violet to red, show us a wealth of detail and the shape of the huge spiral arms which wind around the centre of the nebula.

When we compare photographs of this object taken about half a century ago with our present photographs, we find no evidence of any changes in the system. This, actually, is only an apparent effect due to the great distance which separates us from that galaxy. In reality if we compare the Andromeda nebula with our own Galaxy and we observe carefully the various parts of the nebula, we detect that the whole matter, namely stars of all types and bright and dark nebulae which compose it, is in continuous motion and in rotation. There is no doubt, therefore, that we have here a very large stellar system which must be at about two million light-years distant from us and that probably it is very similar both in size and structure to the stellar system in which our solar system is contained. Do other stellar systems exist and if so how far in space can we find them?

When we observe along the plane of our Galaxy, on one side we see dense clusters of stars embedded in bright or dark cosmic clouds. Even with the naked eye, on a clear night in our regions of the Earth, we can see a wonderful spectacle in the direction of Sagittarius where we believe the centre of the Galaxy to be. If we turn round and look along the same plane but in an opposite direction, we can still see a crowding of stars but not as many as when looking towards Sagittarius. When we turn our sight at right angles to the galactic plane towards what we could call its poles, we notice that the stars are not so crowded and we can then see further into space. The giant telescope of Mt. Palomar tells us not only that in that direction there are fewer stars but that there are countless faint points of light which are ill defined and which are not stars but stellar systems or galaxies, which appear so small and faint because of their enormous distance from us. Those which are relatively nearer, show some kind of spiral form. Some of them appear to show just the beginning of such a form while others are much further developed. Within the distance which can be reached by the 200-inch reflector, we can count thousands of millions of these galaxies. What is there beyond this distance? Our soundings are incomplete and we console ourselves by saying that the universe is infinite. Some mathematicians suggest that the universe is finite, but this will not mean much to our senses until such a time when our observations can reach to the limit of the universe if such a limit exists.

2 · The Galaxies

Modern astronomers are more inclined to study celestial objects which are very distant from us rather than those much closer which belong to our solar system. The reason is probably due to the fact that they now have telescopes available which can probe the depths of space.

The continued study of remote objects is of great importance. Today we can assert with confidence that we know much about the structure, the dimensions and the motion of our Galaxy. We are part of the solar system situated near the edge of the large disc which forms the Galaxy and we take part in its general rotation. The rotation takes place around the centre which is situated in that large luminous mass of stars forming the constellations of Ophiucus and Sagittarius, clearly visible towards the south in our hemisphere, in the early hours of a summer evening.

It was no easy task for the astronomer to acquire knowledge of the Galaxy since everything in the sky is in motion and fixed reference points are not available. In addition there was also the difficulty of determining the distance of the most remote celestial objects. Finally we must remember that we are part of this stellar system and at first it did not seem possible that we could obtain a clear knowledge of it as we participate in its motion. Fortunately nature has granted man the opportunity of studying the Galaxy as if he were able to observe it from outside.

We have seen (page 233) that in Andromeda there is a galaxy, the Andromeda nebula, barely visible to the naked eye. However, with the help of photographs obtained with large telescopes, we can see that the nebula consists of a mass, elliptical in shape, brighter at the centre and around which are found spiral arms composed of dense groups of stars and of matter consisting of gas and dust. Such an object was assumed to be a stellar system similar to our Galaxy and, like the Galaxy, to be in motion.

In order to be able to compare these two stellar systems it was essential to know, at least approximately, the constitution and dimension of our own Galaxy and of the galaxy in Andromeda and above all, the distance separating the two. This was an extremely difficult task since there are no direct methods of measuring distances larger than a few hundreds of light-years, and it is necessary to rely on methods which may be less accurate. After many years of observations and research, it was found, with reasonable approximation, that the galaxy in Andromeda is about two million light-years away from the solar system—one light-year corresponds to approximately six million million miles—and that both its dimension and its constitution are very similar to those of our own Galaxy. It has also been possible to determine the rotation of the two systems around their respective centres as well as the sense of rotation, which indicates a winding of the spiral arms around the centre. From this it follows that if we wish to give to our own Galaxy the name of 'universe' then there is at least another galaxy which has the same right to the name.

The relative nearness of the Andromeda galaxy allows us to study its very remarkable details. Perhaps it may seem peculiar, to say the least, to consider two million light-years as being relatively near, but it does not seem so strange when in exploring the universe further, we discover that we can reach, in our exploration, distances which are much greater. We find an infinite number of 'universes' which used to be called 'island universes' in order not to confuse them with the universe as a whole.

The general picture which is offered to us by the photographic exploration of the sky is truly wonderful. Many people are familiar with the beautiful black and white and coloured photographs recently obtained (page 207) of hundreds of celestial objects. Among these we may recall the photograph of a spiral galaxy in which we see a bright nucleus surrounded by spiral arms which seem to wind around the centre. These spiral arms consist of stars and luminous masses, some of which cannot be resolved into stars, and other bright and dark masses of gas and dust. This is the appearance of the stellar system if it is seen face-on from the Earth. However, these systems can have any orientation in the sky. When we see them edgeways they appear as an elongated luminous band, thicker at the centre and often having a dark band lying along the main plane. Between the two extreme cases of face-on and edgeways there is every possible variation, so that we find in the sky, as far as the 200-inch telescope can reach, a great profusion of these objects appearing at all angles. We must add that there are also countless systems which do not show a spiral form, but appear instead globular in shape. These show the characteristic lack of dark matter and no doubt are composed of a great many stars which we cannot resolve because of the distance of the objects.

All these stellar systems, whether spiral or globular, have apparent dimensions and a brightness which may cover a wide range. This is of course due both to their distance from us and to their actual dimensions. Just as in the case of stars, we have here also giant and dwarf systems. The Galaxy and the Andromeda nebula are giant systems. The greatest diameter of these giants is estimated to be about 100,000 light-years, so that when we compare this dimension with the millions and thousands of millions of light-years which express their distances, we realize how far the various systems are from each other.

It is natural that astronomers should ask themselves whether, as in the case of the stars, it is possible to advance any theories concerning the birth, evolution and decay of such systems which must have been created in the past or are continuing to be born in the universe. Many theories have been put forward on this particular question, but we cannot say that any plausible or satisfactory hypothesis has so far been formulated. More observations and investigations are still required and perhaps radioastronomy will be able to help to extend further the exploration of the universe.

We have been able to classify stars into young and old, according to their dimensions, composition and temperature and we can perhaps suggest some hypotheses concerning the stellar systems. The dust and gas, mainly hydrogen, condense at first into a mass of irregular form which begins to rotate in a process similar to the formation of cyclones in our own atmosphere. At a certain point two arms are ejected from the nucleus, and as a result of either a violent explosion or differential rotation, they take the shape of a spiral (fig. 11). It is in these arms that are born the young stars which are very hot.

Fig. 11. Possible evolution of galaxies

As time progresses, the arms gradually wind around the nucleus and when the hydrogen, which is the most powerful element of all celestial life, is exhausted, a spherical shape develops with no trace of spiral arms and dark matter. At this stage only old stars are left which have a much lower temperature. On the other hand, it may be suggested that a continuous evolution from one type to a final type does not take place. The masses of gas which take part in the process and the electromagnetic phenomena which develop, produce the various forms which we observe, until hydrogen, which has given them life, is completely exhausted.

The question is still an open one and without doubt it is one of the most fascinating of all questions facing modern astronomy.

3 · Recent Research on the Galaxy

The Milky Way, that luminous band which majestically crosses the sky and embraces both the poles of the Earth, is a magnificent sight. Looking north we see it crossing Cassiopeia and Cygnus where it divides into two branches, one directed towards Lyra, Ophiucus and Scorpio and the other towards Aquila and Sagittarius. Here stars and cosmic matter are so bright that the sky is completely illuminated and is a beautiful sight. After these constellations, the Milky Way extends into the southern hemisphere, where it can be followed through Centaurus, the Southern Cross and Argo, reaching Canis Major and returning into the northern hemisphere in Canis Minor and part of Orion.

Galileo was the first to give an explanation of the Milky Way. In his *Sidereus Nuncius* in 1610, he wrote: 'In fact the Milky Way is nothing more than a congeries of numerous stars, scattered in groups.'

Two centuries elapsed before William Herschel, who from being a musician became an outstanding astronomer, determined the general form of the Milky Way by studying its stars in the various regions, with the help of a powerful telescope which he had made himself. With good approximation he described the Galaxy as looking like a 'millstone' with a thickness which was one-fifth of its diameter.

As the Earth is inside the Galaxy, astronomers are in an unfavourable position to view the Galaxy as a whole. Nevertheless they put forward tentative hypotheses as more became known about the shape and dimensions of the stellar system. Herschel's comparison of the Galaxy to a millstone, changed in time to a comparison with a biconvex lens, with stars crowding its main plane but then thinning out gradually above and below the plane. Although we cannot state that the stellar system has well-defined limits, because stars and diffuse matter become less dense as they extend into space, we can estimate its diameter to be about 100,000 light-years with an average thickness of about 4,000 light-years, which increases at the centre to about 15,000 light-years. The solar system is situated approximately on the main plane of the 'lens', that is to say on the galactic equator at a distance of about 30,000 light-years from the centre of the Galaxy which, from the Earth, is seen in the direction of Sagittarius.

Imagination was given full rein on what would be the general appearance of the Galaxy to an outside observer at a great distance from it. No man will ever be able to travel outside the Galaxy because of the enormous distances involved and the length of time which would be required. If we imagine that we could leave the Galaxy from its north pole at right angles to the equator, namely moving towards Coma Berenicis and travel to a distance of several thousand light-years, how would our stellar system appear to us? Considering its shape, its dimensions, its composition of stars and of both bright and dark matter, it would appear to us like one of the many galaxies which show beautiful spirals winding around their bright nucleus, in other words like the Andromeda nebula and that in Ursa Major, when seen from the Earth.

The Galaxy would then be a spiral galaxy, like a gigantic Catherine-wheel, of which there are countless examples scattered in the universe. In order to accept this description, it was necessary to discover the existence of spiral arms and to determine the motion of the Galaxy, since we know that no celestial object can be at rest. In any case, the shape of the galaxies suggested that they must rotate around their centres, just as the planets revolve around the Sun. From our place of observation, outside the Galaxy, we would be able to discover the velocity and the sense of rotation of the system, if we had the patience of carrying out measurements for thousands and thousands of years. Since this is only possible in our imagination, it was necessary to find the means of obtaining the same results without leaving the Earth. This became possible a few years ago thanks to the work of a few astronomers, among whom we must mention Oort, director of the Leyden Observatory.

By investigating the motion of stars of various types and at various distances from the solar system, Oort established that the whole system must rotate around a centre situated in Sagittarius where we can see the brightest part of the Milky Way. The rotation, however, was not that of a solid body, that is to say not like that of a wheel, but rather followed, to a certain extent, the laws regulating the motion in the solar system. This meant that the stars which are nearer to the centre of the Galaxy, must move faster than those further away. Let us remember that in our own solar system, Mercury completes one revolution around the Sun in 88 days while Pluto, the furthest of the planets, takes 248 years to complete its revolution around the Sun. Naturally the whole solar system participates in the rotation and performs a complete revolution around the centre of the Galaxy in about 240 million years. This means that the solar system and all the stars in its neighbourhood travel with a velocity of about 155 miles per second.

It was much more difficult to establish the existence and the position of the spiral arms which surround the nucleus of the Galaxy. The existence of the spiral arms was surmised a few years ago from the distribution of the whitish-blue stars which are very hot and which are found in the more external parts of the Galaxy. At the same time, by means of radio telescopes, it was discovered that neutral hydrogen, scattered in the Galaxy in dark masses, emitted radio waves on a wavelength of 21 centimetres. The hydrogen masses are in motion just like the stars. If they approach the solar system they emit radio waves of a slightly shorter wavelength, while if they recede, the wavelength of their radio waves is slightly longer. Oort and his associates, patiently observing with their radio telescopes, were able to establish where the masses of hydrogen were situated and how they were distributed in the spiral arms. They were also able to determine that these arms are gradually winding around the nucleus. In this way it was proved that our Galaxy is very similar to the myriads of galaxies which we can observe as far as the power of our telescopes permits.

4 · Hydrogen in the Universe

About half a century ago, when spectroscopic analysis of the light of the Sun and of the stars became possible, it was discovered that hydrogen was one of the main components of celestial bodies. Hydrogen is the lightest of chemical elements in the universe. It is found all over the Earth in combination with carbon, oxygen and many other elements. In the conditions in which it exists in the interior of the stars it can produce an enormous amount of energy. This has been proved, by artificial means, even on the Earth. It is the energy from hydrogen which gives life to the Sun and consequently indirectly is responsible for the development and maintenance of life on the Earth. Hydrogen is the main source of life of all the stars as it is found in great abundance both within and outside them.

Only a few years ago it was discovered that the quantity of hydrogen contained in the Sun and in the majority of the stars was much greater than that of all other elements taken together. The spectroscope shows the atoms of hydrogen and its molecular compounds on the surface which is the only part of the star which can be directly investigated. The atoms of hydrogen when they are excited by the very high temperature prevailing on the stars, emit or absorb light at given wavelengths, just as a string of a given length, when plucked, emits always a given musical note.

It has been found that hydrogen represents more than 80 per cent of the volume of the Sun, helium only 18 per cent and the

244

remaining 2 per cent consists of almost all the other known chemical elements, although, of course, in extremely small quantities. The existence of such a large quantity of hydrogen ensures the life of the Sun for thousands of millions of years. Because of the very high temperature and pressure existing in the interior of the Sun, hydrogen is transmuted into helium, the gas which next follows it in the Periodic Table and which received its name because it was first found in the Sun, before its existence on the Earth was discovered. In all the stars we detect the presence of hydrogen even if in varying quantities.

A new discovery was then made. William Herschel, over a century and a half ago, with the aid of his great reflector of 48-inch aperture, noticed that between the two branches of the Milky Way there appeared to be dark regions as if stars were absent there. He was so struck by the existence of these very large dark regions, that he mentioned to his faithful assistant, his sister Caroline, that there appeared to be 'real holes in the sky' through which infinite space could be seen, and he drew the limits of these regions. Only years later did Father Secchi put forward the hypothesis that these dark 'holes' were in fact dark matter scattered between the star clouds of the Milky Way or in the spiral arms of the galaxies like that in Andromeda. This hypothesis was soon amply confirmed by celestial photographs. Nowadays we know that very large clouds of gas are intermingled with the stars. Some of these clouds are more dense than others and are found not only in the interstellar space of the Galaxy and of other galaxies, but probably also in the whole of space, which, until the evidence supplied by the celestial photographs, had been considered empty.

As well as photographing these clouds we can also now determine their composition from their absorption of the light of the stars. The further a star is from the solar system, the longer is the journey of its light through space in order to reach us. Hence the light must travel through a considerable thickness of gas, which, although extremely rarefied, nevertheless imprints its existence on the light of the star which reaches us.

The spectroscopic analysis of the light of the stars has shown that the matter interposed between the stars and the Earth must be composed of atoms of elements in a gaseous state such as calcium, sodium, iron, hydrogen, carbon and nitrogen as well as of molecular compounds. Together with these atoms and molecules there must also be present protons and free electrons, in other words ionized atoms of hydrogen. Recently experimental proof has been obtained by means of radio astronomical observations that, in interstellar matter also, hydrogen is the most important component. Theory predicted that in the conditions in which hydrogen exists in the Galaxy, this gas ought to emit radio waves of a wavelength of 21 centimetres. The theoretical predictions were fully verified by Dutch and American astrophysicists a few years ago. Then a systematic investigation was begun in order to discover the presence of dark clouds of hydrogen in the various parts of the Galaxy.

These investigations carried out on a large scale at the Leyden, Harvard, Washington and Sydney observatories, yielded remarkable results by detecting not only the position of the hydrogen clouds in the neighbourhood of the equatorial plane of our Galaxy, but also their motion. We already know that this large stellar system which extends in space for thousands of light-years, rotates around its own centre of symmetry, like a big Catherine-wheel, not as a solid body but with a velocity which gradually decreases as we proceed outwards from its centre.

The multitude of stars which form the Galaxy are found around its centre in large spiral arms and the dark matter, particularly hydrogen, is distributed mostly in its equatorial plane, between the various spiral arms. The hydrogen masses also take part in the general rotation of the Galaxy and the observations obtained by radio telescopes on a wavelength of 21 centimetres, showed not only their presence by the intensity of the radio waves which are received, but also succeeded in establishing the velocity of rotation which, as we have already said, decreases gradually as we move away from the centre.

In the bright regions of the Galaxy, from Orion to Taurus, Auriga, Perseus, Cassiopeia, Cepheus and Cygnus, photographs and the spectroscope show the existence of many stars which are

246

very hot. From their apparent magnitude it has been established that they can be divided into two groups. The first consists of stars at a relatively small distance from us, between 7 and 2,000 light-years and the second consists of stars which are beyond 6,500 light-years. These two groups can be observed in wide regions of the Galaxy and they can be thought to form two separate rings of stars. By analogy with other galaxies where we can actually see the luminous spiral arms interposed with spiral arms of dark matter, it has been suggested that the two rings are nothing more than the two spiral arms of our Galaxy.

Radioastronomy is far superior to optical astronomy in the study of the spiral structure of our Galaxy because it can explore much greater distances. In fact the radio waves, or at least those of a wavelength around 21 centimetres, can travel undisturbed through the cosmic gas and dust which form the dark clouds interposed between the bright clouds, so that the greatest part, if not the whole of the Galaxy, will be explored in due course.

New observations carried out by radio telescopes on a wavelength of 21 centimetres in the direction of the centre of the Galaxy, have shown the presence also in these regions of the ubiquitous hydrogen. Here the gas has a relatively low temperature and is endowed with very turbulent motions which are localized in the central nucleus of the Galaxy. Optically the hydrogen masses cannot be detected because of the bright star clouds which form the constellations of Scorpio, Ophiucus and Sagittarius.

5 · X Rays from the Sky

We have already mentioned that our atmosphere acts as a protective shield to animal and vegetable life on the Earth by stopping the harmful part of the powerful solar radiation. At the same time the atmosphere prevents man from studying the Sun and the stars in detail because it absorbs a great part of the electromagnetic radiations emitted by them. In addition the atmosphere's continuous turbulence introduces distortions in the images. The rapid progress of space research allows us at present to explore the sky beyond the limit of our atmosphere. In effect we can now build orbiting observatories which are equipped with instruments suitable for various investigations.

Our atmosphere is transparent to those radiations which can be detected both by our eyes and by our photographic plates, namely the radiations which give rise to light. In addition it is also transparent to radio waves of wavelengths ranging from a centimetre to a few metres. On the other hand it is completely opaque to radiations of higher frequencies such as those of the far ultraviolet, X rays and γ rays. Recently rockets, space probes and artificial satellites carrying suitable instruments and capable of rising above the terrestrial atmosphere, investigated for the first time this new domain of radiation.

The γ rays are electromagnetic radiations of a much higher frequency than light waves and X rays. They do not carry electric charges and therefore they can be considered as particles electrically neutral. The X rays, which are so well known to us because of their medical applications, are also high-frequency radiations, but their frequency is lower than that of γ rays. They are more penetrating than light rays since they can travel through obstacles which stop light. The γ rays on the other hand are even more penetrating than X rays and can travel through lead of a thickness of a few inches, so that rays emitted by radium can be detected even after travelling through lead 12 inches thick.

There exists a great selection of instruments which enables the recording of electromagnetic radiations of various frequencies. Radio waves of centimetre wavelength can be received on the Earth by radio telescopes of various types and can be recorded on paper by pen recording instruments. The radiations of the visible spectrum are usually photographed by means of spectrographs, that is by instruments which record on a photographic plate the image of the spectrum from violet to red. Spectrographs can be carried by rockets and recovered by means of parachutes or can be carried by artificial satellites and space probes controlled from the Earth. The γ and X rays can be recorded by means of counters installed in artificial satellites acting as observatories such as O.S.O. (orbiting solar observatory). An O.S.O. satellite in 1963 performed over a thousand orbits around the Earth and transmitted to the Earth measurements of the intensity of γ rays.

Rapid progress in the technique of radioastronomy has for some years enabled us to detect a large number of radio sources scattered in the sky, and which very often do not seem to coincide with any visible celestial object. It is only recently, however, that it has become possible to record the emission of X rays from celestial sources which have been identified. In 1963 an Aerobee rocket, equipped with a powerful X ray detector, was launched by the United States. When the rocket reached heights between 40 and 125 miles the detector explored the whole of the sky during its four minutes of flight and detected an intense source of X rays in Scorpio and another, less strong, which coincided with the Crab Nebula in Taurus. Between two stars belonging to Scorpio, one of which is Antares, the equipment recorded the presence of intense X rays in a region of the sky covering nearly 2 degrees, which is equal to four times the diameter of the Moon, without there being any indication of a luminous source.

It was assumed that as in the case of the Sun, these X ray sources would also be radio sources emitting radiations of frequencies which could be detected by radio telescopes, but this was not the case.

For the time being, the only theories that can be put forward assume that we are dealing here either with an invisible cloud of very hot neutral gas, or with a neutron star. It is not the first time that the theory of the existence of neutron stars has been advanced. In studying the phenomena shown by supernovae (page 214) it has been suggested that when they collapse, stars are formed which consist only of neutrons and which have a density that is extremely high by our terrestrial standards. Such stars would be so faint as not to be observable visually and could be only discovered by the presence of X rays.

The other source of X rays whose intensity is only about one-tenth that in Scorpio, coincides with the Crab nebula. Astronomers, who have studied this exceptional Crab nebula in every detail, are convinced that it is the remnant of a supernova which appeared in 1054 and which was so bright as to be visible even in daylight and hence it must have been brighter than Venus. At present, bright gas surrounds a very faint star, which, because of its character, may well represent the remnants of the explosion of the supernova. The X rays reaching us from this nebula, could be produced by a thermal emission which is still active, 912 years after the explosion. This in the astronomical scale of time represents only a very short time indeed.

Besides the well defined sources of X rays, a diffuse flux of X rays has also been recorded in the sky as a whole, but it is not possible to establish whether this is of galactic or extragalactic origin. It would be of great importance to know the answer to this particular question, in connection with the discussions which are at present taking place concerning the evolution of the universe.

Many other similar experiments are being carried out and

others are being prepared for the study of X rays from space. This has opened a new field of research which, together with radioastronomy, will no doubt add to our knowledge of the 'invisible' sky.

6 · The Visible and Invisible Universe

The progress of research in the last few years, largely due to the advance of modern technology, has made us aware that there is a world much more complex and varied beyond the limitation of our senses.

The great celestial atlas, the Sky Atlas, compiled after many years of work with one of the powerful telescopes of Mt. Palomar, shows us that the various regions of the sky are very different if we compare photographs taken with blue sensitive emulsions with those taken with red sensitive emulsions. In other words, if our eyes were sensitive only to blue light we would see the starry sky very differently from what we would see if our eyes were sensitive only to red light. Stars which appear very faint in blue light, appear very bright in red light. Masses of gas, both bright and dark, are projected on the sky among stars of one colour or another, with shapes which often are very different. What has been said for these two colours in particular, is also true for other colours, and the appearance of the sky will be even more unusual when it is possible to photograph it, as has already been attempted, from great heights above our atmosphere which absorbs all the ultraviolet radiations.

The spectrum of the Sun is seen and photographed by us through the terrestrial atmosphere, but outside this, as has already been shown by rockets which have reached several miles above the Earth, its extent is much greater and remarkable discoveries have already followed concerning the constitution of the Sun. Applying the same principles to the stars, we have a completely new field to explore.

The other field which is just as wide and even more promising, is that of radioastronomy. Notwithstanding the great development of this branch of astronomy in a very short time, we must recognize that we are only at its beginning. Large numbers of physicists and of radio engineers are involved in this work as well as a few astronomers.

It is not surprising that the first to be interested in this new branch of astronomy and who gave great impetus to its development, were physicists who had specialized in electronics. After all the equipment which is suitable for receiving radio waves from space, is very similar to the equipment used in normal radio communications. Many of the radar installations built and used during World War II, could easily be adapted for research in radioastronomy.

Since it was discovered, about 30 years ago, that it was possible to receive radio waves from outer space, radio telescopes of increasing size were designed and built. Radio telescopes have no optical parts of course, but in their most common form consist of a large concave bowl made of metal, having at its centre an aerial which captures electromagnetic waves and feeds them into a receiver to which a pen recorder is connected, in order to obtain a permanent record. These large bowls or reflectors are mounted in the same way as astronomical telescopes so that they can be directed towards any part of the sky. They have to be of large dimensions in order to have a good resolving power. Thus a very close partnership has developed between astronomers and electronic engineers in the investigation of celestial radio sources which, as we have already explained (page 246), has led to new discoveries about the composition of the universe.

Radio telescopes of various shape, size and power are nowadays active all over the world. Some are only engaged in the investigations concerning the radio emissions from the Sun, while others are engaged in the much greater task of searching the whole sky for objects visible and invisible which emit radio waves. The discovery of radio emission from the Sun, at various wavelengths from the centimetre to the decametre wave-band, has been relatively easy. When the Sun is very active, as happened in 1947 and in the period 1957–1958 when it was even more active, very intense radiation was detected. The radio waves emitted in these circumstances enabled us to discover how solar phenomena are propagated. The violent eruptions of hydrogen travel from the photosphere, as the lowest levels of the solar atmosphere are called, to higher levels and reach distances which are equal to many radii of the Sun, namely to the corona. We can actually consider these eruptions as real radio storms which develop in the corona, sometimes with a period of between 5 and 10 days. The electromagnetic bursts which develop in these cases follow each other with great frequency and if our eyes were sensitive to wavelengths between 1 and 15 metres, they would be dazzled by an increase in brightness of the Sun equivalent to hundreds of times that of its normal value. We can also measure the velocity with which these bursts rise in the solar atmosphere, and we find that it increases gradually from 300 to 1,900 miles per second. Such velocities are, on average, those with which the corpuscles emitted by the Sun during a storm, reach our atmosphere after a day or two, and produce such phenomena as aurorae and magnetic storms.

With the progress achieved in radio communications, it was thought that one day it would be possible to send signals to the nearest planets such as Venus and Mars, and even receive transmissions from these planets if intelligent beings existed on them, who had mastered the technique of electronics. As we have already mentioned earlier, contact with the planets of the solar system, particularly Jupiter and Saturn, has been established, but it is only one way traffic. There is no doubt that the emissions we receive from these planets are natural phenomena, the result of electromagnetic disturbances which develop on the planets and which are similar to some of the phenomena occurring on the Earth. No doubt the Earth itself emits into space radio waves of various intensities which could be detected by other planets if the suitable equipment existed on them.

We have already discussed how from visual and spectroscopic

observations we can reach the conclusion that it is highly improbable that human beings like ourselves exist on the other planets of the solar system. Certainly that is the case for Jupiter and Saturn which are liquid and gaseous bodies. If human beings existed on Mars or Venus, it would not be difficult to distinguish between what we could call natural and artificial emission of radio waves from the difference which would exist between the pattern of the signals.

We can easily appreciate what a wide field of research lies ahead for radio telescopes outside the solar system, both in the region of our Galaxy as well as outside it, into the greatest depths of space. The most important fact which has already emerged is that generally speaking there are very few sources of radio waves which coincide with visible celestial objects. We already know that in the whole of space which was thought to be empty, there actually exists a great quantity of interstellar matter consisting mainly of hydrogen. Visual and photographic observations have revealed the existence of dark or bright clouds according to whether in the neighbourhood stars exist which are capable of exciting the gas or not. Unfortunately neither visual nor photographic observations can penetrate either type of clouds and so our knowledge of their extension and motion is limited. The main difference between optical and radio observations is that while optical radiation is stopped by these clouds, radio waves are not. Radio waves can travel through interstellar matter and reveal to us regions of our Galaxy which until now have been completely unknown because they were hidden by the clouds of cosmic dust and gas. Radio astronomers have thus been able to establish the fact that the interstellar gas is particularly dense in the arms of the spiral of the Galaxy and that, together with all the stars, it is in rotation around the centre of the Galaxy. In the direction of this centre, which had already been established by optical means, a radio source has been discovered, which is situated in Sagittarius and which is believed to be the nucleus of the Galaxy. This nucleus, which is very bright and compact and has a diameter of 2,000 light-years, must be surrounded by a cloud of hydrogen which, according to the observations made by the radio telescopes, must be in continuous expansion.

Very few other radio sources in the Galaxy coincide with visible objects. All the stars, many of which are larger and hotter than the Sun, are bound to emit radio waves, but they are so far away from us that the signals must be heavily attenuated and cannot be detected by our instruments.

Outside the Galaxy we know of the existence in large numbers of galaxies some of which have dimensions comparable to those of our Galaxy. These also must be radio sources but only in a few cases has it been possible to detect the radio waves emitted. Radio sources which have been identified with visible objects have only been detected with the help of powerful optical instruments such as the 200-inch reflector of Mt. Palomar. Generally these sources do not consist of a single stellar system. In one case photographs revealed that the radio source coincides with two spiral galaxies which are in contact with each other. Their two nuclei appear distorted in the photographs either on account of the reciprocal gravitational attraction or on account of other processes which we must admit we do not understand, at least at present. It is very unlikely that in these systems actual collisions of stars take place with consequent release of enormous energies, because we know that as the stars generally are so far apart from each other, their encounter is very improbable. The suggestion which has tentatively been advanced is that the clouds of dust and gas existing in interstellar space, meet in their rapid and turbulent motion so that a very large kinetical energy is developed which is capable of emitting intense radio waves.

The name of Jodrell Bank has been often mentioned in recent years in connection with the tracking of artificial satellites. It is located near Manchester and is part of Manchester University. It is the site where Sir Bernard Lovell erected a giant radio telescope which has been active since 1957 and is by now well known the world over. It is still one of the largest radio telescopes in existence which is steerable and can be directed to any region of the sky visible from Jodrell Bank. The radio telescope consists of a large parabolic reflector made of steel. Its diameter is 250 feet and at the centre it has a depth of 60 feet. The aerial which receives the radio waves from the sky is situated at the focus of the reflector. The instrument is mounted equatorially like an optical telescope and is driven so as to be able to follow the apparent diurnal motion of the stars in the sky. The aerial is connected to a receiver and to a pen recorder which can record the radio emission of any particular celestial object (Plate 6).

A radio telescope of this size was originally designed and used by Lovell for astronomical research. With the increase in the number of launchings of artificial satellites in recent years, the powerful and flexible instrument at Jodrell Bank was called upon to help in tracking the artificial satellites particularly before the transmitters installed in them began their automatic transmissions. Apart from this secondary duty the Jodrell Bank radio telescope has, in the few years of its life, accomplished important astronomical discoveries. It has searched for radio sources, for giant stars, nebulae, galaxies and many other celestial bodies which cannot be detected visually. The instrument has also been widely used in the study of meteors and meteor showers, a study which is so important for our knowledge of the upper atmosphere, particularly in connection with the development of space travel.

The trails left in the sky by meteors consist of ionized gases which can be detected by radio telescopes when the trails themselves are intercepted by a beam of electromagnetic waves emitted from a transmitter on the Earth. Jodrell Bank and other centres in the world follow systematically these observations the results of which give the information necessary to determine the distance, the velocity, the frequency and the density of meteors as well as the direction from which they come.

Let us get back now to celestial objects of larger dimensions, the study of which already occupies many observatories all over the world. The main purpose of the investigations is to discover the origin and nature of the radio waves emitted at various frequencies and with various intensities. Already much has been discovered concerning the star nearest to us, that is the Sun, which because of its nearness sends to us very intense radio waves and hence can be investigated in much greater detail than any other star. The Sun-storms, which can be observed both visually and photographically, correspond in many ways to the radio emissions. Novae and supernovae also emit radio waves which we can receive with our radio telescopes. However, when we leave our Galaxy and search for radio sources at distances of the order of millions and thousands of millions of light-years, we must have recourse to much more powerful and more sensitive radio telescopes if we wish to record the flux whose intensity is indirectly proportional to the distance of the sources.

A few years ago Sir Martin Ryle, another well-known radio astronomer, at Cambridge, designed and built a radio telescope which is based on the principle of an interferometer of great sensitivity. By means of this instrument he was able to probe space at a distance much greater than that reached by the 200-inch optical telescope of Mt. Palomar. At these distances, which are of the order of a few thousand million light-years, galaxies are observed which recede from us with a velocity which increases in proportion to their distance. It is the phenomenon which is generally accepted as a proof of the expansion of the universe. On this subject theoretical astronomers have attempted to formulate theories which try to explain the origin of the universe. At present there are two fashionable theories, one known as the 'evolution theory' and the other as the 'steady-state theory'. Lovell, in a course of lectures broadcast not long ago by the B.B.C., explained in a simple and extremely clear manner the main principles upon which both theories rest. The important question is to prove experimentally which of the two theories is more probably correct. The evolution theory suggests that the universe originated thousands of millions of years ago in the explosion of an extremely dense nucleus, and that matter scattered in space by the explosion gradually condensed into nebulae and star clusters. The steady-state theory considers rather a creation which is continuous and always in evolution, and that stars are formed from the basic element which is hydrogen. If the evolution theory is correct then there should be a crowding of galaxies at the extreme limits of the universe. On the other hand the steady-state theory would imply no such crowding of galaxies but rather that the density should be the same in all regions of space.

The preliminary investigations of Ryle and his associates seem to show that at a distance of about 8,000 million light-years from the Earth, the flux of radio waves is about 10 times more intense than in the neighbourhood of the solar system and this would, for the time being at least, lead us to accept the evolution theory.

Optical and radio telescopes complement each other and integrate their work of the study of the universe and at the same time compete in reaching the furthest possible distances of the universe—which appears to be limitless. The 200-inch reflector can photograph galaxies up to a distance of a few thousand million light-years. Recently photographic plates have been replaced by photoelectric devices and amplifiers which can reveal the existence of celestial objects so faint that their light could not be sufficient to affect photographic emulsions. By these means the power of the telescope can be greatly increased and from the first experiments it seems that distances of the order of 2,000 million light-years can be reached. Let us remember that light travels at a speed of 186,000 miles per second and that there are 86,400 seconds in one day alone!

Much greater distances can be reached with radio telescopes because while it is extremely difficult to build an optical mirror of a diameter larger than 200 inches, it is relatively easy to build metallic bowls of much greater dimensions. These are required not only to collect the weak flux of radio waves received from remote celestial objects, but also to detect and to separate individual sources, so that we can obtain their accurate position in the sky. From experiments already carried out, we know that in many cases radio telescopes are the only instruments which can reveal to us the existence of sources which are not visible.

With instruments of increasing power it will be possible to determine with accuracy the distances of the planets of the solar system. Already by using radar astronomy we can, with our present radio telescopes, transmit pulses to the Sun and planets and receive back the echo. Since the velocity of propagation of the pulses is known, being equal to that of light, we can measure the interval of time between the emission of the pulse and the reception on the Earth of the reflected signal and hence determine, with great accuracy, the distance of the celestial body in question. In the case of the Sun which, as we already know, is surrounded by extensive envelopes of gas such as the chromosphere and the corona, the time taken by the signals on their return journey will give us some information about the depth of the various layers through which they travel, or will give us some details of the particular regions where solar phenomena occur. Perhaps it will be even possible, in due course, to record the signals reflected by the clouds of particles which during certain phenomena are ejected from the Sun. Radar astronomy has been used to determine the distance of Venus.

Radio telescopes have already given us indications of the temperature prevailing on the surface of Venus, of Mars and of Jupiter, because the temperature can be obtained from the density of the flux of radio waves emitted by the planets. From measurements made on the centimetre wave-band, Venus appears to have a surface temperature of 330°C. while the temperature measured in the infrared region of the spectrum is about −25°C.

These widely differing results may well indicate that the high temperatures which are measured are non-thermal in origin. They are perhaps produced by electrons which are accelerated by intense magnetic fields to a very high velocity, approaching that of light.

7 · New Discoveries at the Horizon of the Universe

In previous chapters we have often mentioned the difficulties which face astronomers in their efforts to correlate radio sources with objects which can be seen and observed by ordinary astronomical methods. No difficulty is experienced in recording radio waves emitted by the Sun, but all the other stars populating our Galaxy are in general too far from us for our radio telescopes to record any emissions which probably exist. If we leave the Galaxy aside, there are all the galaxies in the universe which are undoubtedly radio sources. Astronomers are making a great effort to find whether any visible celestial object coincides with any particular radio source. Nowadays the theory of the expansion of the universe is generally accepted and we know that the very distant galaxies appear to recede from us with a velocity which is proportional to their distance. It is obvious that whether we can determine such velocities of recession as far as the limit of the universe, if a limit exists, depends on the power of our instruments as well as on the velocity of the galaxies under consideration, because if the velocity is equal to that of light then the galaxies become invisible to us.

Before the co-operation between radioastronomy and optical astronomy had started, the 200-inch reflector of Mt. Palomar had succeeded in photographing galaxies which were at a distance of approximately 2,000 million light-years. From the red-shift which is apparent in their spectra, calculations indicate that such galaxies would have a velocity of recession equal to half the velocity of light, namely 93,000 miles per second.

A couple of years ago, radio telescopes discovered radio sources which had an apparent stellar structure to which the name 'Quasars' (Quasi-Stellar Objects) was given. Soon a dozen or so of these objects were identified and appeared to be among the brightest and most remote objects which we could investigate in the universe. It is calculated that an average quasar emits as much energy as a hundred galaxies. The investigation of the spectrum of some of these objects by means of the 200-inch reflector, revealed that the spectrum is similar to that of other galaxies, showing a red-shift which followed the usual law in relation to the distance of the object emitting the spectrum. It seems, therefore, that quasars are galaxies which have dimensions from 10 to 100 times smaller than ordinary galaxies. One of them, among the furthest of all, appears to recede from us with a velocity of nearly 150,000 miles per second, which is approximately 80 per cent of the velocity of light.

An even greater surprise, however, was in store for the astronomer. Recently the 200-inch reflector, independently of any radio telescopes, showed the existence of another class of objects similar in appearance to quasars but which did not emit radio waves. These objects were discovered photographically on account of their excessive blue colour which indicated extremely high temperatures, so that they were named 'blue stellar objects'. It has been established that these objects recede from us with a velocity which is as high as that of the quasars and hence they too must be very near to the limit of the range of our instruments. Blue stellar objects seem to exist in great numbers and to be furthest away from us, thereby adding considerable weight to the theory that the expansion of the universe was caused by an initial explosion. The problem which these objects present is that, although they possess enormous energy, as is shown by their colour, they do not emit radio waves. In order to avoid confusion with the quasars, these objects have been named 'Quasi-Stellar Galaxies' (Q.S.G.). Their silence in the domain of radio could be due to the nature of the matter of which they were composed at the time of their formation. If they were composed only of neutrons, which are elementary particles devoid of any electric charge, the lack of radio emission could then be easily explained. No answer is as yet available to explain this mysterious new class of objects.

All these new objects are seen as they were a few thousand million years ago, that is to say very near to the time when the universe was formed, which, according to the theory of the expanding universe, must be of the order of 10,000 to 15,000 million years ago. This discovery may lead us to assume that we are actually seeing galaxies in the process of formation, revealing what was the structure of galaxies in that far distant time. Moreover, if in those remote regions of the universe we detected a slowing down of the velocity of expansion of the universe, then we would have some evidence for one of the theories which suggests that the universe, following an expansion may then contract again, thus undergoing a great and periodic pulsation. The crowding of galaxies in the most distant regions of the universe together with the enormous release of energy by quasars and by Q.S.G.s would confirm the hypothesis of an original explosion. We must remember that the solar system, at the time of its origin, could not have been at the centre of the primeval nucleus from which the universe was formed. All around us we can see galaxies of all types, from spiral to elliptical, and therefore we are in the presence of objects which must have very different ages.

The mysteries of the universe are still numerous and we must be patient and hope to unravel them slowly, one at a time.

8 · The Unity of the Universe

Up to a century ago our forefathers could determine the brightness, colour and motion of celestial objects simply by observing them. What they could not do was to acquire a knowledge of the matter composing these bodies, nor could they even guess that, in a not too distant future, it would have been possible to solve such an important problem.

From the great amount of heat and energy emitted by the Sun, they assumed that it was an incandescent sphere, but it was not clear how it could burn for so long without consuming itself. In the case of the stars, their distance from the Earth was completely unknown and, although it was accepted that stars shone with their own light like the Sun, nothing was known about their real dimensions.

Galileo realized from his early observations that the planets of the solar system were dark bodies, which shone because of reflected light, but at the time astronomers lacked the techniques to establish whether or not planets had an atmosphere. Only in the case of the Moon could they suggest that it had no atmosphere because their observations showed an unchangeable surface, devoid of clouds.

A discovery of great importance was made in the last century which enabled astronomers to study celestial bodies from a physical point of view. A method was discovered to detect the composition of the Sun and of the stars, and in the case of the planets, to detect the composition of their atmosphere.

The wonderful natural phenomenon known as the 'rainbow' shows how, in some particular circumstances, the white light of the Sun can be refracted by drops of rain and give rise to a series of colours from blue to red, producing the solar spectrum. The same phenomenon can be produced artificially by letting a beam of sunlight travel through a glass prism. In analysing the spectrum of the Sun or that of the stars, it became possible to determine the substances of which the celestial bodies were composed.

Spectroscopes, added to our powerful telescopes, enable us to observe visually or to obtain photographs of the spectra of the Sun and stars. The study of these spectra gives us all the information we may require about the stars' outer layers which are extremely hot, so hot in fact that all the chemical elements known on the Earth in a solid or liquid state, appear in the spectra of the stars in a gaseous state. The atoms of those gases which are very highly excited by the high temperatures prevailing on the celestial bodies, emit radiations in various regions of the spectrum, from the ultraviolet to the infrared. The elements can be identified simply by comparing the spectra of celestial bodies with the spectra produced in our laboratories, of the various elements, at least as far as we can reproduce in our terrestrial laboratories the conditions existing on the celestial bodies.

Spectroscopic analysis has led to a great discovery: the Sun is composed of the same matter which on the Earth we find in a solid, liquid or gaseous state. The same applies to the stars, both for those which have similar temperature to that of the Sun and for those which have higher or lower temperatures. In the nebulae scattered in our Galaxy, in all the bodies which form the Galaxy and indeed even in the numerous remote galaxies, matter is always the same, thus proving the unity of the universe.

On the Earth, ninety-two simple and natural elements have been discovered, from the lightest which is hydrogen and is in a gaseous state, to the most complex and heavy which is uranium. Almost all these elements, although in different physical conditions and quantities, have been identified in the atmosphere of the Sun and stars. In fact by studying stellar spectra, astronomers can make an analysis both qualitative and quantitative of the more external envelopes of the stars. A comparison between the composition of the Sun and that of the Earth shows that in the case of the lighter elements such as hydrogen, which is extremely abundant on the Sun, and oxygen, the composition of the Sun is very different from that of the Earth, while in the case of elements heavier than oxygen, the chemical composition of the Sun can be considered very similar to that of the terrestrial crust and of meteorites.

The light of the Sun which reaches and penetrates the atmosphere of the planets is then reflected to the Earth. Investigation of the spectrum of this reflected light shows that the composition of the atmosphere of the planets consists of gases existing on our Earth which are well known to us. Thus the light from Venus and Mars shows the existence of carbon dioxide, a small amount of oxygen and very little hydrogen. The light of the greater planets, Jupiter, Saturn, Uranus and Neptune reveals large quantities of methane and ammonia. There is very little doubt that the internal constitution also of all the planets of the solar system must be similar to that of the Earth.

Just as there is a unity of matter in the universe, which we can investigate, so there is in existence a unity of physical forces dominating the universe. Material bodies fall on the Earth because of the force of gravity which the Earth exerts upon them. Likewise the Moon revolves around the Earth because of the same force, as Newton proved, and the same law is valid for all the planets of the solar system. It was a short step from this to discover, from the motion of double stars and the motion of stars around the centre of our Galaxy, that the law of gravitation, which is valid for the Earth and the Sun, is the same everywhere in the universe. The study of the magnetic fields which exist on the Sun and on the stars and in general, of the electromagnetic phenomena which can be observed in the whole of space, indicates that these forces are everywhere the same, even if in space they develop to a scale so great that it is beyond our comprehension.

266

9 · Life in the Universe. Invisible Solar Systems

Everything in the universe is in motion, everything is alive. We cannot conceive stars as being stationary and lifeless in space. By 'life' we mean a continuous and powerful production of energy which originates from the transmutation of the matter of which the stars are composed. The planets of the solar system, and possibly those in other similar systems which are themselves dark, still have a life of a similar type which develops either in their interior or on their surface or externally in their atmosphere. Even the Moon, which we know to be devoid of an atmosphere, may have an internal life of its own, since the matter of which it is composed could conceivably be radioactive and hence release energy.

On the other hand if we mean by 'life' animal and vegetable life such as that which has developed on the Earth, we must admit that it is almost only on the Earth that we can detect it and study it. We have said 'almost' because there seems to be some evidence of the presence of a vegetable life in the solar system, in the form of a very sparse vegetation, perhaps something like our lichens, on the surface of Mars. The recent information obtained by Mariner II seems to exclude the possibility of life similar to ours on Venus because of the high temperature of the dense clouds which surround it.

The Copernican revolution demoted the Earth from its position as the centre of the universe to that of an ordinary planet moving around the Sun. It is strange that today we seem to have reached the conclusion that among all the planets of the solar system it is only on the Earth that a fully developed animal and vegetable life can be found. This consideration is very important if we think of the unity of the physical forces which rule the universe, because even limiting ourselves to the solar system it would appear reasonable to think that on the other planets there should be at least some form of life similar to ours. This question of existence of life is not an easy one to answer if we think of the very different physical conditions of the various planets of the solar system and their relation to the central star, the Sun, which is the giver of light and heat. These physical conditions can be summarized in the dimensions of the planets and in the values of gravity on their surface, and in the composition and temperature of their atmosphere.

Can man be so arrogant today as to assert or believe that he is the only inhabitant not only of the solar system but of the whole universe?

To this question astronomy gives an answer which, though at present vague and hypothetical, may perhaps well become more certain in the not too distant future.

Today there is strong evidence that other solar systems exist. These cannot be seen but can be discovered mathematically. This assertion may require some explanation.

Celestial bodies, like all other material bodies, exert an attraction according to the well-known Newtonian law. The larger the mass of a body, the greater is its attraction, which is inversely proportional to the square of the distance. When we are dealing with only two celestial bodies in motion, and all celestial bodies are moving, it is easy to calculate their motion around their common centre of gravity. We have a typical example in the case of binary stars. With prolonged observations we can see the two stars which form the system, revolving around each other.

Better still if we take as reference one of the two stars, generally the brighter, we can see that the companion describes an orbit which is generally elliptical. Celestial mechanics teaches us how, from this apparent orbit, we can calculate the real orbit in space, described in this case by the companion around the main star.

Observations have shown that in the sky there are as well as binary systems also systems consisting of three or more stars. In addition we also have the case of the solar system which consists of a bright star of considerable mass, surrounded by a family of planets which, in their turn, are surrounded by satellites. We are faced here with a situation where we are dealing with more than two bodies, and the individual attraction of each becomes confused and complicates considerably the motion of each system. Once again we turn to celestial mechanics for help. Unfortunately the problem of three bodies is so complex that a general solution has not yet been found. In the majority of cases, celestial mechanics can only offer us an approximate solution. This is obtained by calculating how any one of the three or more celestial bodies which form the system, is subject to 'perturbations' by the presence of the others, and from this knowledge we can determine its position in the past, present and future. Nowadays electronic computers simplify considerably the calculations which in the past were so lengthy and laborious. In the case of the solar system, the mass of the Sun is so much greater than the sum of the masses of all its planets that in a first approximation the effect of the reciprocal attraction of the planets can be ignored. However, when we wish to obtain very accurate results, such as those which give the position of the planets during the year and which are published in the astronomical tables, then we must take the perturbations into account. Jupiter, the largest of the planets of the solar system, is the cause of considerable perturbations in the motion of the other planets. When Jupiter approaches the Earth its attraction slows down the motion of the Earth in its orbit, and when Jupiter recedes the Earth accelerates in its orbit.

It is natural to think that our solar system is not unique in the universe and to speculate that at least some of the many stars existing in the universe could be accompanied by planets. It does not necessarily follow, however, that the conditions prevailing on the Earth exist in these other systems. The circumstances which are so vital for the evolution of animal and vegetable life on the Earth are briefly as follows: distance from the Sun, the dimension of the Earth and the quantity of electromagnetic radiation which it receives. Of all the planets of the solar system only the Earth seems to have these favourable conditions. Nevertheless considering the enormous number of stars which exist in the Galaxy, many of which are of the same type as our Sun, we must admit the probability of the existence of some planets at least endowed with characteristics similar to those prevailing on the planets of our own solar system. Even if such planets existed and were as large as Jupiter, they would still be invisible to us with our present methods of observation. We can only attempt to determine their existence, therefore, by indirect means based on the theory of perturbations.

It is a well-known fact that the discovery of Neptune, and of Pluto years later, was first made by theoretical means. It was a real triumph for celestial mechanics, because both planets were easily located in the sky, following the instructions of the astronomers who had calculated the perturbations produced by these unknown bodies in the orbits of the other planets of the solar system.

There is no hope, in the case of single stars, of discovering whether they have any planets because we cannot determine the perturbations of their motion. In the case of binary systems the problem is easier.

For two binary systems, 61 Cygni and 70 Ophiuchi, it has been possible to determine the existence of planets which accompany one or other of their components.

61 Cygni is a system which is relatively near to us. Its distance is approximately 11 light-years and it consists of two components which are 'red dwarfs' so called because their mass is only half that of the Sun and their temperature is lower than that of the Sun. Observations carried out over a number of years have shown that the two stars revolve around each other with a period of

720 years. The distance between the two components is nearly 110 times the distance between the Earth and the Sun. The apparent orbit of one component with reference to the other should be an ellipse but actually periodic perturbations have been detected which reveal the existence of a third body which must have a mass 16 times that of Jupiter. This third body must complete a revolution around one of the two components in about five years, and must be at an average distance from it

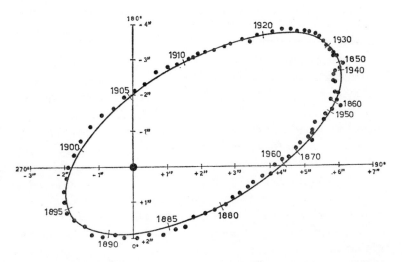

Fig. 12. Irregularities of the orbit of the fainter component of 70 Ophiuchi

equal to twice the distance between the Earth and the Sun.

70 Ophiuchi is a binary system which was discovered in 1825 and has been regularly observed since then. One of the two components of the system is fainter than the other and revolves around its brighter companion with a period of 88 years, so that its apparent orbit with reference to the brighter companion is now well known (fig. 12). The two stars forming this system are red dwarfs. The larger has a mass almost equal to that of our Sun, while the mass of the companion is less than half that of the Sun. The distance of the system from us is approximately 16 light-years. The fainter star, instead of describing a regular orbit

271

around its companion, is subject to oscillations. During a whole revolution these oscillations periodically take it in and out of the regular ellipse which it ought to describe in the absence of perturbations. The perturbations can be explained by the presence of a third very small body, but we cannot tell around which of the two stars this third body is moving. The period of its revolution is 17 years and its mass is nearly ten times that of Jupiter. Because of its very small dimensions, this body must be dark and hence it may well be a planet which is invisible to us.

Other binary systems similar to the two described above have been discovered and therefore we may reasonably conclude that our Sun is certainly not the only star which has the privilege of being accompanied by a family of planets.

Assuming that other planets exist beyond the solar system, will man ever be able to establish communications with possible beings which may sooner or later inhabit such planets? As we have mentioned earlier in this book, radio waves have already been transmitted from the Earth and have bounced back from the Moon and Venus, but these are celestial bodies which are very close to us when we think in terms of interstellar distances. If we were to send a radio message to the star 61 Cygni, we would have to wait 22 years before we could receive a reply to our message from any inhabitants, if they existed, of the planets of that star.

From the technical point of view there do not seem to be serious difficulties in the sending of radio waves of suitable wavelength, even to such enormous distances, but how could our signals be understood?

Already radio signals have been sent out from the Green Bank Observatory in West Virginia, to stars which it is thought may have a system of planets. The stars chosen are at a distance of about 10 light-years from us, so we still have a few years to wait for a possible reply. Of course we could receive radio signals sent to us many years ago by hypothetical inhabitants of other planets if they had conceived the idea that the star which we call the Sun was accompanied by planets on which life existed.

Man on the Earth is patiently scanning space and listening in;

encouraged by the success already achieved. Who can foretell the answer to many of the problems which we have discussed in this book? We live in an age of fantastic and exciting progress which may well lead us to a better knowledge of the universe around us.

Index

275